SCIENCE
AND THE
AGELESS
WISDOM

BOOKS BY ALICE A. BAILEY

Initiation, Human and Solar
Letters on Occult Meditation
The Consciousness of the Atom
A Treatise on Cosmic Fire
The Light of the Soul
The Soul and its Mechanism
From Intellect to Intuition
A Treatise on White Magic
From Bethlehem to Calvary
Discipleship in the New Age–Vol. I
Discipleship in the New Age–Vol. II
Problems of Humanity
The Reappearance of the Christ
The Destiny of the Nations
Glamour: A World Problem
Telepathy and the Etheric Vehicle
The Unfinished Autobiography
Education in the New Age
The Externalisation of the Hierarchy

A Treatise on the Seven Rays:
Vol. I – Esoteric Psychology
Vol. II – Esoteric Psychology
Vol. III – Esoteric Astrology
Vol. IV – Esoteric Healing
Vol. V – The Rays and the Initiations

Master Index of the books of Alice. A. Bailey
The Labours of Hercules: An Astrological Interpretation

COPYRIGHT © 2025 BY LUCIS TRUST

First Printed Edition, 2025

ISBN 978-085330-147-9

All rights reserved. No part of this book may be reproduced or utilised in any form or by any means, electronic or mechanical, including photocopying, recording, or by any information storage and retrieval system, without permission in writing from the publisher.

The non-profit Lucis Publishing Companies are owned by the Lucis Trust. As the original publishers of all of the books of esoteric philosophy by Alice A. Bailey, the companies are dedicated to ensuring the ongoing availability, authenticity, and production quality of the publications. No royalties are paid and publications are financed by the Tibetan Book Fund which has been established by Lucis Trust for the perpetuation of the teachings of the Tibetan and Alice A. Bailey.

Science and the Ageless Wisdom
is also available in eBook format

LUCIS PUBLISHING COMPANY
866 United Nations Plaza, Suite 482, New York, NY 10017

LUCIS PRESS LIMITED
Suite 54, 3 Whitehall Court, London SW1A 2EF

www.lucistrust.org

PRINTED IN THE UNITED KINGDOM

SCIENCE
AND THE
AGELESS WISDOM

*From the Writings of
Alice A. Bailey
and
the Tibetan Master, Djwhal Khul*

LUCIS PUBLISHING COMPANIES

you to ascertain their truth by right practice and by the exercise of the intuition. Neither I nor A.A.B. is the least interested in having them acclaimed as inspired writings, or in having anyone speak of them (with bated breath) as being the work of one of the Masters. If they present truth in such a way that it follows sequentially upon that already offered in the world teachings, if the information given raises the aspiration and the will-to-serve from the plane of the emotions to that of the mind (the plane whereon the Masters *can* be found) then they will have served their purpose. If the teaching conveyed calls forth a response from the illumined mind of the worker in the world, and brings a flashing forth of his intuition, then let that teaching be accepted. But not otherwise. If the statements meet with eventual corroboration, or are deemed true under the test of the Law of Correspondences, then that is well and good. But should this not be so, let not the student accept what is said.

AUGUST 1934

EXTRACT FROM A STATEMENT
BY THE TIBETAN

Suffice it to say, that I am a Tibetan disciple of a certain degree, and this tells you but little, for all are disciples from the humblest aspirant up to, and beyond, the Christ Himself. I live in a physical body like other men, on the borders of Tibet, and at times (from the exoteric standpoint) preside over a large group of Tibetan lamas, when my other duties permit. It is this fact that has caused it to be reported that I am an abbot of this particular lamasery. Those associated with me in the work of the Hierarchy (and all true disciples are associated in this work) know me by still another name and office. A.A.B. knows who I am and recognises me by two of my names.

I am a brother of yours, who has travelled a little longer upon the Path than has the average student, and has therefore incurred greater responsibilities. I am one who has wrestled and fought his way into a greater measure of light than has the aspirant who will read this article, and I must therefore act as a transmitter of the light, no matter what the cost. I am not an old man, as age counts among the teachers, yet I am not young or inexperienced. My work is to teach and spread the knowledge of the Ageless Wisdom wherever I can find a response, and I have been doing this for many years. I seek also to help the Master M. and the Master K.H. whenever opportunity offers, for I have been long connected with Them and with Their work. In all the above, I have told you much; yet at the same time I have told you nothing which would lead you to offer me that blind obedience and the foolish devotion which the emotional aspirant offers to the Guru and Master whom he is as yet unable to contact. Nor will he make that desired contact until he has transmuted emotional devotion into unselfish service to humanity, – not to the Master.

The books that I have written are sent out with no claim for their acceptance. They may, or may not, be correct, true and useful. It is for

THE GREAT INVOCATION

From the point of Light within the Mind of God
    Let light stream forth into the minds of men.
    Let Light descend on Earth.

From the point of Love within the Heart of God
    Let love stream forth into the hearts of men.
    May Christ return to Earth.

From the centre where the Will of God is known
    Let purpose guide the little wills of men—
    The purpose which the Masters know and serve.

From the centre which we call the race of men
    Let the Plan of Love and Light work out.
    And may it seal the door where evil dwells.

Let Light and Love and Power restore the Plan on Earth.

"The above Invocation or Prayer does not belong to any person or group but to all Humanity. The beauty and the strength of this Invocation lies in its simplicity, and in its expression of certain central truths which all men, innately and normally, accept—the truth of the existence of a basic Intelligence to Whom we vaguely give the name of God; the truth that behind all outer seeming, the motivating power of the universe is Love; the truth that a great Individuality came to earth, called by Christians, the Christ, and embodied that love so that we could understand; the truth that both love and intelligence are effects of what is called the Will of God; and finally the self-evident truth that only through *humanity* itself can the Divine Plan work out."

ALICE A. BAILEY

Contents

INTRODUCTION	1
I. SCIENCE AND ESOTERICISM	3
II. SPIRIT – LIGHT – MATTER:	7
The Origin of Consciousness	7
III. HYLOZOISM	10
IV. COSMOLOGY	12
Chart V. Evolution of A Solar Logos	18
V. ELECTRICITY – THE BASIS OF UNIVERSALITY	19
VI. THE SEVEN PLANES AND THE SEVEN RAYS	30
1. Science and the Planes	31
Chart III. Seven Planes of our Solar System.	32
2. Science and the Rays	43
a. Cosmic Rays	52
VII. COSMIC AND SYSTEMIC LAWS	55
VIII. CREATION – THE BUILDING OF FORM	58
1. Cosmic Fire–The Origin of Energy	63
2. Etheric Matter	67
a. Etheric Centres	75
3. Prana	77
4. Life	83
5. Evolution–Involution	87
6. Consciousness–Mind–Manas	90
7. The Soul	96
IX. THE GREATER AND LESSER BUILDERS	102
1. The Entifying of the Solar Sphere	102
Diagram of The Atom	106
a. The Hierarchies	109
2. Agni–The Ruler of Fire	111
3. The Greater Builders–Fire Devas of the Physical Plane	114
4. The Transmitting Devas	116
5. The Lesser Builders–The Fire Elementals	117

a. Physical Plane Elementals	118
1) Elementals of Densest Matter	118
2) Elementals of Liquid Matter	123
3) Elementals of Gaseous Matter	126
4) Elementals of the Ethers	128
6. Man–A Builder in Mental Matter	131
Chart VIII – Cosmic Physical Plane	134
7. The Human Soul and the Permanent Atoms	135
8. Human and Deva Evolution Compared	146
X. CELESTIAL DYNAMICS	151
1. Constellations	151
2. The Solar System	155
3. The Sun	158
4. Planetoids–Asteroids	161
5. Planet Earth	163
a. Appropriation by Planetary Logos of Physical Body	170
6. Moon Chain	173
7. Milky Way	175
XI. STATES OF MATTER	177
1. Atoms–Electrons–Ions	177
2. Solid–Matter and Substance	181
3. Liquid	184
4. Gaseous	185
XII. ATOMIC ENERGY	187
1. Predictions	191
XIII. PHYSICAL FORCES OF EVOLUTION	194
1. Sound–Light–Vibration–Form	194
2. Sound	196
a. Communication	203
b. Music of the Spheres	204
3. Light	206
4. Vibration	211
5. Colour	214

a. Forecasts on Colour, Sound and Vibration	216
6. Magnetism	218
7. Polarity	225
8. Gravitation	228
9. Heat	230
10. Motion	233
a. Rotary	235
b. Spiral-Cyclic	237
c. Forward Progressive	239
11. Energy and Force	240
12. Attraction–Repulsion–Cohesion	246
13. Expansion	254
14. Space and Time	256
XIV. RADIATION AND TRANSMUTATION	265
1. Transmutation in the Human Kingdom	276
2. Transmutation in the Mineral Kingdom	277
3. Alchemy	279
XV. IMMORTALITY	285

This compilation is extracted from the books by Alice A. Bailey.

The compiler has introduced titles and headers into the chapters/quotations.

Throughout the text Helena P Blavatsky author of *The Secret Doctrine* is referred to as H.P.B.

INTRODUCTION

This book contains a structured collection of excerpts from the Alice Bailey writings mainly focussed on science, yet also including topics related to philosophy and religion. Human knowledge and thinking is evolving all the time. Scientists or philosophers of the middle Ages or the time of the Enlightenment could hardly conceive the scientific view of today. And who knows what the scientific view of the world will be in a century, or even half a century from now?

This book was compiled with two types of readers in mind:
1. Scientists with an open mind looking for possible extensions of their current viewpoint.
2. Esoteric students wanting to know more about the scientific implications of their esoteric studies.

AS A SCIENTIST:

When reading this book as a scientist, a prerequisite is an open mind. Many of the suggestions made are beyond the current scientific viewpoint summarised as the Standard Model and the Big Bang theory. The Ageless Wisdom holds that in the universe, all is electric in nature and this is a concept always to be kept in mind, even though we have to acknowledge that we still do not know the true nature of what electricity is. Scientists with a physics or mathematical background are advised to extend their comprehension of the concepts of electricity and magnetism beyond the limiting ideas as expressed by the equations of Maxwell. In the Ageless Wisdom, electricity and magnetism have a deeper, profounder meaning. Other concepts of science, when viewed from the angle of the Ageless Wisdom are often endowed with a deeper meaning too.

A point to keep in mind too is that the separation of living and dead matter from the point of view of science, does not hold in the Ageless Wisdom: In the Ageless Wisdom the whole range from the mineral, vegetable, animal to the human kingdoms are considered to be alive, as well as the planets, solar systems and galaxies.

At first reading, some of the excerpts may appear abstruse or even incomprehensible. This is mainly because the Ageless Wisdom is not limited to our materialistic concept of a three dimensional world in space with time as a 'fourth dimension'. The Ageless Wisdom is based on a much richer cosmology with many concepts borrowed from the Eastern philosophies. Our Western, scientific cosmology is only a tiny part of this far more complete cosmology, including both visible and invisible worlds. Once the reader can embrace these wider concepts – where nothing is left to chance – a rich but very structured picture emerges. It is then that some of the excerpts of this book may spark new insights or new ideas.

AS AN ESOTERIC STUDENT:

Readers who are already somewhat familiar with the concepts of the Ageless Wisdom will find in the excerpts many examples of links to the world of science. Some of these links point to the future, some to the past or present state of scientific affairs. Starting with the wider concepts of the Ageless Wisdom, it will be seen how science is slowly forging its path forward to a more complete understanding of the universe in all its expressions: ranging from human beings to the planets and the cosmos, but also downward to the world of atoms and molecules.

Several concepts in the excerpts refer to events or developments in the future. Some of these developments have not (yet) taken place, which does not mean that they will not take place at all. Time is always a complex factor as time on the inner planes is not. Human and planetary evolution are not carved in rock and are always subject to many interdependent factors, thus making predictions extremely difficult and sometimes imprecise.

The goal of this book is not to give a complete picture, neither of the Ageless Wisdom nor of the various concepts of the sciences of today. That is obviously impossible. Independent of the reader's background, it is hoped that this compilation of excerpts may spark new insights, new ideas and understanding.

I. SCIENCE AND ESOTERICISM

• When the Knowledge of God shall shine forth universally (and this is not the knowledge of, or awareness of a great Being but the expression through human instrumentality of the divine omniscience), then will the Lord of Concrete Science, Who is the embodiment of the fifth principle of mind, see His work brought to a conclusion. He stimulates the sense of awareness in humanity and nurtures the consciousness aspect in the subhuman kingdoms, producing the response, therefore, of matter to spirit, and bringing about the interpretation of that to which there has been a sentient rapport.

Esoteric Psychology I, p. 134

• Much of the true revelation since the time of Christ has come to the world along the line of science. The presentation, for instance, of material substance (scientifically proven) as essentially only a form of energy was as great a revelation as any given by the Christ or the Buddha. It completely revolutionised men's thinking and was—little as you may think it—a major blow struck at the Great Illusion. It related energy to force, form to life, and man to God and held the secret of transformation, transmutation and transfiguration.

Glamour: A World Problem, p. 187

• Applied science in all fields has now been so developed that it has entered the realm of energy and of pure metaphysics. The study of matter has landed us in the realm of mysticism and of transcendentalism. Science and Religion are joining hands in the world of the unseen and intangible. *From Intellect to Intuition,* p. 203

• *The Science of Electricity*. The investigations of scientists have been greatly stimulated by the discovery of radium, which is an electrical phenomenon of a certain kind, and by the knowledge this discovery brought of the radioactive substances; the development of the many methods of utilising electricity has also greatly aided. This science has brought man to the threshold of a discovery which will revolutionise world thought on these matters, and which will eventually solve a great part of the economic problem, thus leaving many more persons free for mental growth and work.

A Treatise on Cosmic Fire, pp. 808/9

• Men are now perhaps ready to penetrate beneath the surface and to carry their search within the outer form of nature to that which is its cause. We are perhaps, too apt to confuse the religious spirit with the mystic search. All clear thinking about life and about the great laws of nature, if carried forward with persistence and steadfastness, leads eventually into the mystic world, and this the foremost scientists of our day are beginning to realize. Religion starts with the accepted hypothesis of the unseen and the mystical. But science arrives at the same point by working from the seen to the unseen and from the objective to the subjective. Thus, as has been said, by the process of investigation and of passing inwards from form to form, the mystic arrives eventually at the glory of the unveiled Self. It seems to be unalterably true that all paths lead to God — viewing God as the ultimate goal, the symbol of man's search for Reality. It is no longer a sign of superstition to believe in a higher dimension and in another world of Being. Even the word "supernatural" has become deeply and profoundly respectable, and it seems possible that some day our educational systems may regard the preparation of the individual to transcend his natural limitations as an entirely legitimate part of its affairs. *From Intellect to Intuition,* pp. 37/8

• H.P.B. prophesied the work now being done many years ago when she spoke of the recognition ultimately to be accorded by science to an universally diffused omnipresent Deity (the ether of space is an entity, she also tells us) and that the mystery of electricity, when solved, holds for us the solution of most of our problems. Many of the theories of modern science are laid down in *A Treatise on Cosmic Fire*, though scientists have not gone far enough yet to recognise this fact; there the electrical nature of man is posited. You would find it interesting and helpful to search out such passages. Science, however, gives no place to the electrical force of the soul, which is steadily growing in potency. A few of the scientists among the most advanced are beginning to do this. The next step ahead for science is the discovery of the soul, a discovery which will revolutionise, though not negate, the majority of their theories.

Esoteric Healing, p. 368

• Generally speaking, science has preceded esotericism in its recognition of energy as a dominant factor in all form expression. Theosophists and others pride themselves on being ahead of human thinking, but such is not the case. H.P.B., an initiate of high standing, presented views ahead of science, but that does not apply to the exponents of the theosophical teaching. The fact of all manifested forms being forms of energy, and that the true human form is no exception, is the gift of science to humanity and not the gift of occultism. The demonstration that light and matter are synonymous terms is also a scientific conclusion. Esotericists have always known this, but their aggressive and foolish presentations of the truth have greatly handicapped the Hierarchy. *Telepathy and the Etheric Vehicle,* p. 140

• The limitation of modern science is its lack of vision; the hope of modern science is that it does recognise truth when proven. Truth in all circumstances is essential and in this matter science gives a desirable lead, even though it ignores and despises occultism. Occult scientists handicap themselves either because of their presentation of the truth or because of a false humility. Both are equally bad.

Telepathy and the Etheric Vehicle, p. 142

• [Published in 1925]: *A group of scientists* will come into incarnation on the physical plane during the next seventy-five years who will be the medium for the revelation of the next three truths concerning electrical phenomena. A formula of truth concerning this aspect of manifestation was prepared by initiates on the fifth Ray at the close of the last century, being part of the usual attempt of the Hierarchy to promote evolutionary development at the close of every cycle of one hundred years. Certain parts (two fifths) of that formula have worked out through the achievements of such men as Edison and those who participate in his type of endeavour, and through the work of those who have dealt with the subject of radium and radioactivity. Three more parts of the same formula have still to come, and will embody all that it is possible or safe for man to know anent the physical plane manifestation of electricity during the fifth subrace.

A Treatise on Cosmic Fire, pp. 455/6

II. SPIRIT – LIGHT – MATTER
The Origin of Consciousness

• Two factors are universally recognised in all systems that merit the name of philosophy; they are the two factors of spirit and matter, of purusha and prakriti. There is at times a tendency to confound such terms as "life and form," "consciousness and the vehicle of consciousness" with the terms "Spirit and matter." They are related, but clarity of view would be facilitated if it were realised that *prior to manifestation*, or to the birth of a solar system, it is more correct to utilise the words, Spirit and matter. When these two are inter-related *during manifestation*, and after the cessation of the pralayic interval or interlude between two systems, then the terms, life and form, consciousness and its vehicles, are more correct, for during the period of abstraction consciousness is not, form is not, and life, demonstrating as an actual principle, is not. There is Spirit-substance but in a state of quiescence, of utter neutrality, of negativity, and of passivity. In manifestation the two are approximated; they interact upon each other; activity supersedes quiescence; positivity replaces negativity; movement is seen in place of passivity, and the two primordial factors are no longer neutral to each other, but attract and repulse, interact and utilise. Then and only then, can we have form animated by life, and consciousness demonstrated through appropriate vehicles.

How can this be expressed? In terms of fire, when the two electric poles are brought into definite relationship we have demonstrated, along the line of occult sight and of occult feeling, both heat and light. This relationship is brought about and perfected during the evolutionary process. This heat and light are produced by the union of the two poles, or by the occult marriage of male and female, of Spirit (father) and matter (mother). In terms of the physical, this union produces the objective solar system, the Son of the Father and the Mother. In terms of the subjective, it produces the Sun, as the sum total of the qualities of light and heat. In terms of fire, by the union or at-one-ment of electric fire (Spirit) and fire by friction (energised matter) solar fire is produced. This solar fire will be distinguished above all else by its evolutionary development, and by the gradual intensification of the heat to be felt, and of the light to be seen.

For a clearer comprehension of this abstract matter, we might consider the microcosm, or man evolving in the three worlds. Man is the product of the approximation (at present imperfect) of the two poles of Spirit (the Father in Heaven) and of matter (the Mother). The result of this union is an individualised Son of God, or unit of the divine Self, an exact replica in miniature on the lowest plane of the great Son of God, the All-Self, who is in Himself the totality of all the miniature sons, of all the individualised Selves, and of each and every unit. The microcosm, expressed in other terms or from the subjective point of view, is a miniature sun distinguished by the qualities of heat and light. At present that light is "under the bushel," or deeply hidden by a veil of matter, but in due process of evolution it will shine forth to such an extent that the veils will be lost from sight in a blaze of exceeding glory. At present the microcosmic heat is of small degree, or the magnetic radiation between the microcosmic units is but little *felt* (in the occult significance of the term), but as time proceeds, the emanations of heat,—due to intensification of the inner flame, coupled with the assimilated radiation of other units—will increase, and become of such proportions that the interaction between the individualised Selves will result in the merging to perfection of the flame within each one, and a blending of the heat; this will proceed until there is "one flame with countless sparks" within it, until the heat is general and balanced. When this is the case and each Son of God is a perfected Sun, characterised by perfectly expressed light and heat, then the entire solar system, the greater Son of God, will be the perfected Sun. *A Treatise on Cosmic Fire,* pp. 227/9

• Under the Law of Magnetic Attraction and owing to the impulsive activity of the Universal Mind as it works out the purposes of the solar Logos or of the planetary Logos these constituents of the matter of space, [the multiplicity of tiny atomic lives] these atoms of substance, are drawn together, manipulated in a rhythmic manner and held together in form. Through this mode of creation, existences come into manifestation, participate in the experience of their particular cycle, whether it is ephemeral, like the life of a butterfly or relatively permanent like the ensouling life of the planetary deity, and vanish. The two aspects concerned, spirit and matter, are brought thus into a close rapport, and necessarily exert an effect upon each other. Matter, so-called, is energised or "lifted up" in the occult sense of the term by its contact with spirit so-called. Spirit, in its turn, is enabled to enhance its vibration through the medium of its experience in matter. The bringing together of these two divine aspects results in the emergence of a third, which we call the soul, and through the medium of the soul, spirit develops a sentiency and a conscious awareness and capacity to respond which remains its permanent possession when the divorce between the two comes around eventually and cyclically.

A Treatise on White Magic, p. 522

III. HYLOZOISM

• . . . I seek to give a brief exegesis of the basic theory of *The Secret Doctrine*, called the hylozoistic theory. This posits a living substance, composed of a multiplicity of sentient lives which are continuously swept into expression by the "breath of the divine Life." This theory recognises no so-called inorganic matter anywhere in the universe, and emphasizes the fact that all forms are built up of infinitesimal lives, which in their totality—great or small—constitute a Life, and that these composite lives, in their turn, are a corporate part of a still greater Life. Thus eventually we have that great scale of lives, manifesting in greater expression and reaching all the way from the tiny life called the atom (with which science deals) up to that vast atomic life which we call a solar system.

This is a briefly and inadequately expressed definition of the doctrine of hylozoism, and is an attempt to interpret and find a meaning in the manifested phenomenal world, with its three main characteristics of life-quality-appearance. Forget not to find the meaning behind all forms and life experiences, and thereby learn to enter into that world of subjective forces which is the true world wherein all occultists work. *Esoteric Psychology* I, p. 149

• The fundamental concept of hylozoism underlies all the esoteric teaching upon the theme of manifesting life. All forms are composed of many forms, and all forms—aggregated or single in nature—are the expression of an indwelling or ensouling life. The fusion of life with living substance produces another aspect of expression: that of consciousness. This consciousness varies according to the natural receptivity of the form, according to its point in evolution, and to its position also in the great chain of Hierarchy.

However, dwarfing every other concept, is the concept of life itself. There is—as far as we have ever been permitted to know—only one Life, expressing itself as Being, as responsive consciousness, and as material appearance. That One Life knows itself (if such a term can be used) as the will-to-be, the will-to-good, and the will-to-know. It will be obvious to you that these are only terms or methods organised to convey a better picture than heretofore.

Telepathy and the Etheric Vehicle, p. 182

•... it remains yet for science to acknowledge the "entified" nature of substance, and thus account for the life that energizes the substance of the three lower subplanes. This recognition by science that all forms are built of intelligent lives will come about when the science of magic begins again to come to the fore, and when the laws of being are better understood. Magic concerns itself with the manipulation of the lesser lives by a greater life; when the scientist begins to work with the consciousness that animates substance (atomic or electronic), and when he brings under his conscious control the forms built of this substance, he will gradually become cognizant of the fact that all entities of all grades and of varying constitutions go to the construction of that which is seen. This will not be until science has definitely admitted the existence of etheric matter as *understood by the occultist*, and until it has developed the hypothesis that this ether is in differing vibrations. When the etheric counterpart of all that exists is allocated to its rightful place, and known to be of more importance in the scale of being than the dense vehicle, being essentially the body of the life, or vitality, then the role of the scientist and the occultist will merge. *A Treatise on Cosmic Fire,* pp. 637/9

IV. COSMOLOGY

• The confines of the Heavens Themselves are illimitable and utterly unknown. Naught but the wildest speculation is possible to the tiny finite minds of men and it profits us not to consider the question. Go out on some clear starlit night and seek to realise that in the many thousands of suns and constellations visible to the unaided eye of man, and in the tens of millions which the modern telescope reveals there is seen the physical manifestation of as many millions of existences; this infers that what is visible is simply those existences who are in incarnation. But only one-seventh of the possible appearances are incarnating. Six-sevenths are out of incarnation, waiting their turn to manifest, and holding back from incarnation until, in the turning of the great wheel, suitable and better conditions may eventuate.

Realise further that the bodies of all these sentient intelligent cosmic, solar and planetary Logoi are constituted of living sentient beings, and the brain reels, and the mind draws back in dismay before such a staggering concept. Yet so it is, and so all moves forward to some unfathomable and magnificent consummation which will only in part begin to be visioned by us when our consciousness has expanded beyond the cosmic physical plane, and beyond the cosmic astral until it can "conceive and think" upon the cosmic mental plane. That supposes a realisation beyond that of the Buddhas who have the consciousness of the cosmic physical plane, and beyond that of the planetary Logoi. It is the consciousness and knowledge of a solar Logos.

To the occult student, who has developed the power of the inner vision, the vault of Heaven can therefore be seen as a blazing fire of light, and the stars as focal points of flame from which radiate streams of dynamic energy. Darkness is light to the illumined Seer, and the secret of the Heavens can be read and expressed in terms of force currents, energy centres, and dynamic fiery systemic peripheries.

A Treatise on Cosmic Fire, pp. 1059/60

• As regards cosmic position, relation and limitation, little can be

COSMOLOGY

said, as e'en to the Heavenly Men Themselves the matter is obscure. That this is necessarily so must be apparent when Their place in the scheme of things is realised and Their relative unimportance is considered. Therefore, we can do no more than accept the fact of the inconceivable magnitude of that EXISTENCE which is manifesting through seven solar systems, and the extension of this concept of Being to embrace the entire vault of the Heavens. It is interesting to bear in mind in this connection that all that is seen, being objective forms or Beings in manifestation through certain spheres of light, may not be all that IS, but that there may lie back of everything visible a vast realm or realms of Existences. The very brain of man reels in contemplation of such a concept. Yet just as there are tens of millions of human beings out of objective manifestation, or discarnate, on the subtler planes of the solar system, so there may be cosmic entities, in rank equal to the ONE ABOUT WHOM NOUGHT MAY BE SAID, Who are in a similar sense discarnate, and found in realms subtler than that of the manifestation of light.

A Treatise on Cosmic Fire, pp. 411/2

• As the system, or the body of the Logos, is carried forward through the energy in all its parts, so is each infinitesimal part speeded on to its similar individual glorification. The many which form the All, and the units which constitute the One, cannot be differentiated as the consummation is achieved. They are merged, and lost in the general "beatific light," as it is sometimes called. We can then extend the concept somewhat further, and realise the cosmic interplay which is likewise being carried forward. We can picture the cosmic stimulation and intensification which proceeds as constellations form the units in the Whole instead of planets or human atoms. Whole suns with their allied systems in their immensity play the part of atoms. Thus some idea may be gained of the unified purpose underlying the turning of the great Wheel of the cosmic Heaven, and the working through of the life purposes of those stupendous Existences Who hold a position in the cosmic Hierarchy similar to that of the "ONE ABOUT WHOM NOUGHT MAY BE SAID,."

A Treatise on Cosmic Fire, pp. 1117/8

• We come now to the consideration of another point, and one of very real moment; it emerges out of what we have been saying anent cycles and is the basis of all periodic phenomena. One of the most elementary of scientific truths is that the earth revolves upon its axis, and that it travels around the sun. One of the truths less recognised, yet withal of equal importance, is that the entire solar system equally revolves upon its axis but in a cycle so vast as to be beyond the powers of ordinary man to comprehend, and which necessitates mathematical formulae of great intricacy. The orbital path of the solar system in the heavens around its cosmic centre is now being sensed, and the general drift also of our constellation is being taken into consideration as a welcome hypothesis. Scientists have not yet admitted into their calculations the fact that our solar system is revolving around a cosmic centre along with six other constellations of even greater magnitude in the majority of cases than ours, only one being approximately of the same magnitude as our solar system. This cosmic centre in turn forms part of a great wheel till—to the eye of the illumined seer—the entire vault of Heaven is seen to be in motion. All the constellations, viewing them as a whole, are impelled in one direction.

The Old Commentary expresses this obscure truth as follows:

"The one wheel turns. One turn alone is made, and every sphere, and suns of all degrees, follow its course. The night of time is lost in it, and kalpas measure less than seconds in the little day of man.

Ten million million kalpas pass, and twice ten million million Brahmic cycles and yet one hour of cosmic time is not completed.

Within the wheel, forming that wheel, are all the lesser wheels from the first to the tenth dimension. These in their cyclic turn hold in their spheres of force other and lesser wheels. Yet many suns compose the cosmic One.

Wheels within wheels, spheres within spheres. Each pursues his course and attracts or rejects his brother, and yet cannot escape from the encircling arms of the mother.

When the wheels of the fourth dimension, of which our sun is one and all that is of lesser force and higher number, such as the eighth and ninth degrees, turn upon themselves, devour each other, and turn and rend their mother, then will the cosmic wheel be ready for a faster revolution."

It will, therefore, be apparent that the power of man to conceive of these whirling constellations, to measure their interaction, and to realise their essential unity is not as yet great enough. We are told that even to the liberated Dhyan Chohan the mystery of that which lies beyond his own solar Ring-Pass-Not is hid.

Certain influences indicate to Him and certain lines of force demonstrate to Him the fact that some constellations are knit with His system in a close and corporate union. We know that the Great Bear, the Pleiades, Draco or the Dragon are in some way associated with the solar system but as yet He knows not their function nor the nature of the other constellations. It must also be remembered that the turning of our tiny systemic wheel, and the revolution of a cosmic wheel can be hastened, or retarded, by influences emanating from unknown or unrealised constellations whose association with a systemic or a cosmic Logos is as mysterious relatively as the effect individuals have upon each other in the human family. This effect is hidden in logoic karma and is beyond the ken of man.

A Treatise on Cosmic Fire, pp. 1083/5

• Cosmically there is a very interesting series of triangles which will be found by the student of esoteric astronomy and of occult cycles. They originate in the central sun of our particular group of solar systems. This series involves the Pleiades. The fact that this is so will not be known until the last decade of the present century, and will not be recognised by science till that time when certain lines of knowledge and discovery will bring scientists to a realisation that there is a third type of electricity, which ever balances and forms the apex of the triangle. But the time is not yet.

A Treatise on Cosmic Fire, p. 664

• In my last instruction I made the statement that meditation was the major creative agent in the universe. There are other universes that are ahead of us in development and, in them, the emphasis may not be upon creation by use of mental energies; others may not be so advanced and, in them, mental energy may be in process of unfolding or expressing itself—in the evolutionary sense. There are also universes and solar systems where the quality and the conditions of the manifesting universe, solar system or planet are unknown to us. It must be borne in mind that though in all manifestations the three aspects (of purpose or will, attraction, magnetic love or plan, and appearance as manifestation of both of these) are necessarily present, the manifesting Entity (responsible for these expressions of divinity) may work through and "occultly declare" conditions and qualities of which we have no experience or knowledge. We may possess utterly no idea in the highest flights of our abstract thinking (and this includes the most advanced thinkers upon our planet) of the nature of the impulses and concepts which animate certain universal Creators.

Discipleship in the New Age II. pp. 207/8

• It is not, however, a profitless task for the disciples and aspirants to catch the dim outline of that structure, that purpose and that destiny which will result from the consummation and fruition of the Plan on earth. It need evoke no sense of futility or of endless striving or of an almost permanent struggle. Given the fact of the finiteness of man and of his life, given the tremendous periphery of the cosmos and the minute nature of our planet, given the vastness of the universe and the realisation that it is but one of countless (literally countless) greater and smaller universes, yet there is present in men and upon our planet a factor and a quality which can enable all these facts to be seen and realised as parts in a whole, and which permits man (escaping, as he can, from his human self-consciousness) to expand his sense of awareness and identity so that the form aspects of life offer no barrier to his all-embracing spirit. It is of use also to write these words and to deal with these ideas, for there are those now coming into incarnation who can and will understand, when present readers are dead and gone. I and you will pass on to other work, but there will be those on earth who can vision the Plan with clarity, and whose vision will be far more inclusive and comprehending than ours. Vision is of the nature of divinity. Expansion is a vital power and prerogative of Deity. Therefore let us struggle to grasp what is possible at our particular stage of development, and leave eternity to reveal its hidden secrets.

Esoteric Psychology II, pp. 219/20

COSMOLOGY

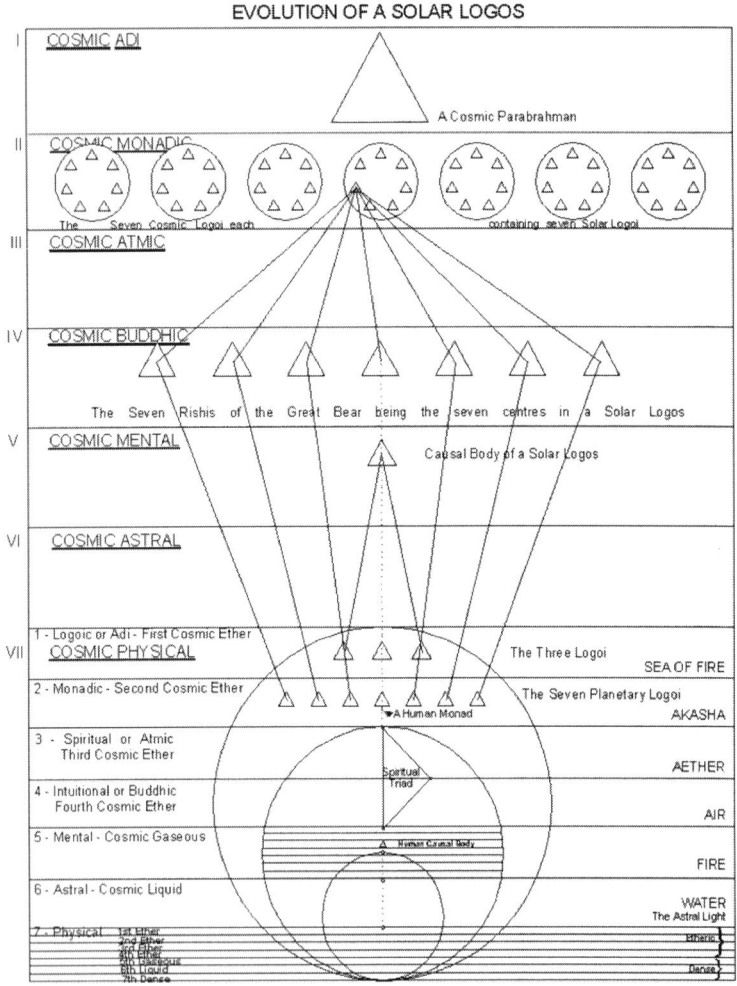

A Treatise on Cosmic Fire, p. 344

V. ELECTRICITY –
THE BASIS OF UNIVERSALITY

• If this treatise serves no other purpose than to direct the attention of the scientific and philosophic students to the study of force or energy in man and in groups, and to interpret man and the human family in terms of electrical phenomena, much good will have been accomplished. The polarity of a man, of a group, and of a congery of groups, the polarity of the planets and their relationship to each other and to the Sun, the polarity of the solar system and its relationship to other systems, the polarity of one plane to another, and of one principle to another, the polarity of the subtler vehicles, and the scientific application of the laws of electricity to the totality of existence on the physical plane will bring about a revolution upon the planet second only to that effected at the time of individualisation.

A Treatise on Cosmic Fire, p. 714

• Whenever the words influence, radiation, or the power of a ray, are used, we are dealing inferentially with electrical phenomena, or with energy of some kind. This energy, or electrical manifestation, this "mystery of electricity" to which H. P. B. refers, is the foundation of all manifestations, and lies back of all evolution. It produces light in ever-growing brilliancy; it builds and moulds the form to the need of the indwelling Entity; it brings about coherence and group activity; it is the warmth that causes all growth, and that fosters not only the manifestations of the vegetable and animal kingdoms but induces interaction between the human units, and lies behind all human relationships. It is magnetism, radiation, attraction and repulsion, life, death, and all things; it is conscious purpose and essential will in objective manifestation, and he who has solved what lies back of electrical phenomena has solved not only the secret of his own Being, but knows his place within his greater sphere, a planetary Logos, is conscious of the Identity of that cosmic Existence we call a solar Logos, and realises somewhat the place of our system and its electrical relationship with the seven constellations.

A Treatise on Cosmic Fire, p. 436

• Man's just apprehension of this mystery of electricity will only come about as he studies himself, and knows himself to be a triple fire, manifesting in many aspects.

A Treatise on Cosmic Fire, p. 608

• It should here be carefully borne in mind that we are dealing with electrical matter, and are therefore concerned with cosmic etheric substance; all matter in the system is necessarily etheric. We are consequently dealing literally with physical phenomena on all planes of the system. In time and space we are concerned with units of different polarity which—during the evolutionary process—seek union, balance, equilibrium or synthesis, and eventually find it. This electrical interplay between two units causes that which we call light, and thereby objectivity. During evolution this demonstrates as heat and magnetic interaction and is the source of all vital growth; at the achievement of the desired goal, at union, or at-one-ment, two things occur:

First, the approximation of the two poles, or their blending, causes a blazing forth, or radiant light.

Second, obscuration, or the final disintegration of matter owing to intense heat.

This can be seen in connection with man, a Heavenly Man and a solar Logos, and their bodies of objectivity. In man this polarity is achieved, the three different types of electrical phenomena are demonstrated, and the light blazes forth, irradiating the causal body, and lighting up the entire sutratma, or thread (literally the Path) which connects the causal vehicle with the physical brain. Then disintegration or destruction ensues; the causal body vanishes in a blaze of electrical fire, and the real "man" or self is abstracted from the three world-bodies. So will it be seen in the body of a Heavenly Man, a planetary scheme, and so likewise in the body of the Logos, a solar system.

ELECTRICITY – THE BASIS OF UNIVERSALITY

The difficulty in apprehending these thoughts is great, for we are necessarily handicapped by lack of adequate terms, but the main ideas only are those I seek to deal with, and the one we are primarily concerned with in this division is the *electrical manifestation of magnetism*, just as earlier we dealt cursorily with the same electrical phenomena, manifesting as the activity of matter.

Therefore you have:

1. Activity………....electrical manifestation of matter.
2. Magnetism........electrical manifestation of form.
3. Vitality………..electrical manifestation of existence.

This is literally (as pointed out by H.P.B) fire by friction, solar fire, and electric fire.

Fire by friction is electricity animating the atoms of matter, or the substance of the solar system, and resulting in:

The spheroidal form of all manifestation.
The innate heat of all spheres.
Differentiation of all atoms one from another.

Solar fire is electricity animating forms or congeries of atoms, and resulting in:

Coherent groups.
The radiation from all groups, or the magnetic interaction of these groups.
The synthesis of form.

Electric fire is electricity demonstrating as vitality or the will-to-be of some Entity, and manifests as:

Abstract Being.
Darkness.
Unity.

We have seen that electrical manifestation on the first plane caused initial vibration, and on the second its activity resulted in the archetypal form of all manifestation from a God to man, and an atom.

On the third plane which is primarily the plane of Brahma, this electrical force showed itself in intelligent purpose. The will-to-be, and the form desired, are correlated by intelligent purpose underlying all.

A Treatise on Cosmic Fire, pp. 314/6

- *Manas is electricity.*

The fire of Mind is fundamentally *electricity*, shown in its higher workings, and not considered so much as force in matter. Electricity in the solar system shows itself in seven major forms, which might be expressed as follows:

Electricity on the first plane, the logoic or divine, demonstrates as the Will-to-be, the primary aspect of that force which eventually results in objectivity. Cosmically considered, it is that initial impulse or vibration, which emanates from the logoic causal body on the cosmic mental plane, and makes contact with the first cosmic etheric, or the solar plane of adi.

Electricity on the monadic plane demonstrates as the first manifestation of form, as that which causes forms to cohere. Matter (electrified by "fire by friction") and the electric fire of spirit meet and blend, and form appears. Form is the result of the desire for existence, hence the dynamic fire of Will is transmuted into the burning fire of Desire. I would call attention to the choice of those two phrases, which might also be expressed under the terms:

Dynamic electrical manifestation.
Burning electrical manifestation.

Here on the second plane, the sea of electrical fire, which distinguished the first plane, is transformed into the akasha, or burning etheric matter. It is the plane of the flaming Sun, just as the first plane is that of the fire mist or the nebulae. This idea will be easier to comprehend if it is borne in mind that we are dealing with the *cosmic physical plane*.

Certain things take place on the second plane which need realisation, even if already theoretically conceded:

Heat or flaming radiation is first seen.

Form is taken, and the spheroidal shape of all existence originates. The first interplay between the polar opposites is felt.

Differentiation is first seen, not only in the recognised duality of all things, but in differentiation in motion; two vibrations are recognised.

Certain vibratory factors begin to work such as attraction, repulsion, discriminative rejection, coherent assimilation, and the allied manifestation of revolving forms, orbital paths and the beginning of that curious downward pull into matter that results in evolution itself.

The primary seven manifestations of logoic existence find expression and the three, with the four, commence their work.

The seven wheels, or etheric centres in the body etheric of that great cosmic Entity, of Whom our solar Logos is a reflection, begin to vibrate and His life activity can be seen.

A Treatise on Cosmic Fire, pp. 310/12

• On the third plane, which is primarily the plane of Brahma, this electrical force showed itself in intelligent purpose. The will-to-be, and the form desired, are correlated by intelligent purpose underlying all. . . . On this third plane that intelligent principle demonstrates as coherent activity, either systemic, planetary, or monadic, and also as the triple vibration of spirit-matter-intelligence, sounding as the threefold Sacred Word, or electricity manifesting as sound.

A Treatise on Cosmic Fire, pp. 316, 318

• *On the fourth plane this electrical force shows itself as colour.* In these four we have the fundamental concepts of all manifestation; all four have an electrical dynamic origin; all are basically a differentiation or effect of impulse, emanating from the cosmic mental plane and taking form (with intelligent purpose in view) on the cosmic physical.

A Treatise on Cosmic Fire, p. 319

• The *third Logos* is fire in matter. He burns by friction, and gains speed and added vibration by the rotation of the spheres, their interplay thus producing friction with each other.

The *second Logos* is solar fire. He is the fire of matter and the electric fire of Spirit blended, producing, in time and space, that fire which we call solar. He is the quality of the flame, or the essential flame, produced by this merging. A correspondence to this may be seen in the radiatory fire of matter, and in the emanation, for instance, from the central sun, from a planet, or from a human being,—which latter emanation we call magnetism. A man's emanation, or characteristic vibration, is the result of the blending of Spirit and matter, and the relative adequacy of the matter, or the form, to the life within. The objective solar system, or the sun in manifestation, is the result of the blending of Spirit (electric fire) with matter (fire by friction), and the emanations of the Son, in time and space, are dependent upon the adequacy of the matter, and of the form to the life within.

The *first Logos* is electric fire, the fire of pure Spirit. Yet in manifestation He is the Son, for by union with matter (the mother) the Son is produced by Whom He is known. "I and my Father are One" (John, 10: 30.) is the most occult statement in the Christian Bible, for it not only refers to the union of a man with his source, the monad, via the ego, but to the union of all life with its source, the will aspect, the first Logos. *A Treatise on Cosmic Fire,* pp. 150/1

• It will be demonstrated later as science attains more and more of the truth that

1. All physical phenomena as we understand the term have an electrical origin, and an initial vibration on the first sub-plane of the physical plane.
2. That Light, physical plane light, has a close connection with, and uses, as a medium, the second ether.
3. That sound functions through the third ether.
4. That colour in a peculiar sense is allied to the fourth ether.

We must note here that in the development of the senses, hearing preceded sight, as sound precedes colour.

A Treatise on Cosmic Fire, pp. 319/20

ELECTRICITY – THE BASIS OF UNIVERSALITY

• "We know of no phenomenon in nature - entirely unconnected with either magnetism or electricity - since, where there are motion, heat, friction, light, there magnetism and its Alter Ego (according to our humble opinion) - electricity will always appear, as either cause or effect - or rather both if we but fathom the manifestation to its origin. All the phenomena of earth currents, terrestrial magnetism and atmospheric electricity, are due to the fact that the earth is an electrified conductor, whose potential is ever changing owing to its rotation and its annual orbital motion, the successive cooling and heating of the air, the formation of clouds and rain, storms and winds, etc. This you may perhaps find in some text book. But then Science would be unwilling to admit that all these changes are due to Akashic magnetism incessantly generating electric currents which tend to restore the disturbed equilibrium." (*Mahatma Letters* to A. P. Sinnett, p. 160)

A Treatise on Cosmic Fire, footnote, pp. 310/11

• . . . *only when the substance* aspect is studied by the scientist in its triple nature will truth be approximated, and the true nature of electrical phenomena be comprehended; then and only then will electricity be harnessed and utilised by man as a unity, and not just in one of its aspects as at present; the negative electricity of the planet is all that is as yet contacted for commercial purposes. It must be remembered that this term is used in the sense of negative in relation to solar electricity.

A Treatise on Cosmic Fire, pp. 523/4

• The following are the three basic mysteries of the solar system:

1. *The mystery of Electricity.* The mystery of Brahma. The secret of the third aspect. It is latent in the physical sun.
2. *The mystery of Polarity*, or of the universal sex impulse. The secret of the second aspect. It is latent in the Heart of the Sun, i.e., in the subjective Sun.
3. *The mystery of Fire* itself, or the dynamic central systemic force. The secret of the first aspect. It is latent in the central spiritual sun. . . .

The *mystery of electricity* has three keys, each of which is held in the hands of one of the Buddhas of Activity. Theirs is the prerogative to control the electrical forces of the physical plane, and Theirs the right to direct the three major streams of this type of force in connection with *our present globe*. These three streams are concerned with atomic substance, out of which all forms are constructed. In connection with *our chain* there are three mysterious Entities (of whom our three Pratyeka Buddhas are but the Earth reflections) Who perform a similar function in connection with the electrical forces of the chain. *In the scheme*, the planetary Logos has also three co-operating Existences Who are the summation of His third Aspect, and who perform therefore work similar to that performed by the three aspects of Brahma in the solar system. The mystery of this threefold type of electricity is largely connected with the lesser Builders, with the elemental essence in one particular aspect,—its lowest and most profound for men to apprehend as it concerns the secret of that which "substands" or "stands back" of all that is objective. In a secondary sense it concerns the forces in the ethers which are those which energise and produce the activities of all atoms. Another type deals with the electrical phenomenon which finds its expression in the light which man has somewhat harnessed, in the phenomena such as thunder storms and the manifestation of lightning, with the aurora borealis, and in the production of earthquakes and all volcanic action. All these manifestations are based on electrical activity of some kind, and have to do with the "soul of things," or with the essence of matter. The old Commentary says:

"The garment of God is driven aside by the energy of His movements, and the real Man stands revealed, yet remains hidden, for who knows the secret of a man as it exists in his own self-recognition."

The mystery of electricity deals with the "garment" of God, just as the mystery of polarity deals with His "form."

In the mystery of Polarity, we have three different types of force manifesting and thus it is apparent that the two mysteries deal with the six forces. These three types of force are manipulated by the Buddhas

ELECTRICITY – THE BASIS OF UNIVERSALITY

of Love. They, through Their sacrifice, concern Themselves with the problem of sex, or of "magnetic approach" on all the planes. The Buddha of Whom we speak and Who contacts His people at the full moon of Wesak, is one of the three connected with *our globe*, having taken the place of One Who passed on to higher work in connection with *the Chain*, for the same hierarchical grading is seen as in connection with the Buddhas of Action. One group might be considered the divine Carpenters of the planetary system, the other the divine Assemblers of its parts and the Ones Who, through the magnetic influence They wield, unite the diversities and build them into form.

The present ideas anent Sex must be transmuted and raised from the existing lower connotation to its true significance. Sex—in the three worlds—has to do with the work of the lunar Pitris and the solar Lords. It signifies essentially the form-building work in substance, and its energising by the spiritual aspect. It signifies the elevation of the material aspect through the influence of Spirit as the two together perform their legitimate function in co-operation and thus—by their mutual union and blending—produce the Son in all His glory. This method of interpreting it is equally true of all the Existences manifesting on any plane, systemic and cosmic. Certain factors enter into the thought of sex which might be enumerated as follows:

   a. Mutual attraction,

   b. Complementary suitability,

   c. Instinctual appeal,

   d. Approach, and recognised co-operation,

   e. Union,

   f. The next stage is the temporary importance of the material aspect, that of the Mother, the feminine aspect,

   g. The withdrawal into a temporary retirement of the Father,

   h. The work of creating the Son,

   i. The evolution and growth of the Son, both materially and in consciousness,

   j. Emancipation of the Son from his Mother, or the liberation of the soul at maturity from matter,

   k. Recognition by the Son of the Father and his return to that Father,

The final result of all these successive stages being that all the three aspects have performed their functions (their dharma) on the physical plane and all three have demonstrated certain types of energy.

The Father aspect manifests in giving the initial impulse or the positive electrical demonstration which is the germ of the created Son, and Whose Life is embodied in the Son. The occult significance of the words of the Christ in answer to the cry "Lord, show us the Father" is little appreciated. "He that has seen Me has seen the Father, for I and my Father are One," He said.

The Mother, or the negative aspect, builds and nourishes, guards and cherishes the Son through the ante-natal, and the infant stages, and stands around Him during later stages, giving of the energy of her own body and activity in ministry to His need.

The Son, the combined energy of Father and of Mother, embodies both types and all the dual sets of qualities, but has a character all His Own, an essence which is His peculiar nature, and an energy which leads Him to fulfil His Own ends and projects, and which will eventually cause Him to repeat the process of producing,—

1. Conception,
2. Creation,
3. Conscious growth,

as did His Father.

When we reach the *mystery of Fire*, we are concerned with that mysterious extra-systemic energy which is the basis of both the activity of Mother and the Life of the Son. The Son in very deed "becomes His Mother's husband," as say the ancient Scriptures. This is but an enigmatical phrase unless interpreted in terms of the combination of energy. Only when the Son has reached maturity and knows Himself as essentially the same as the Father can He consciously perform His Father's function, and produce and perpetuate that which is needed for the sustaining of cosmic generation.

The electricity of substance, the electricity of form, and the electricity of Life itself must blend and meet before the true Man (whether

ELECTRICITY – THE BASIS OF UNIVERSALITY

Logos, or human being) realises himself as creator. Man at this stage knows somewhat of the electricity of substance, and is coming to the belief as to the electricity of form (even though as yet he calls it magnetism) but as yet he knows nought of the electrical reality of life itself. Only when the "jewel in the Lotus" is about to be revealed, or the third circle of petals is about to open up, does the initiate begin to have a realisation of the true meaning of the word "life" or spirit. The consciousness has to be fully awakened before he can ever understand that great energising something of which the other types of energy are but expressions. *A Treatise on Cosmic Fire,* pp. 872/6

• We might now take the four minor Rays, which, with the third, make the sumtotal of manas, and see wherein their influence may be expected. The subject is so stupendous that we cannot possibly do more than touch upon certain points, nor can we enlarge along the line of the mechanistic development of forms to utilise the force. This is all hidden in the science of electricity, and as exoteric science discovers how:

> To utilise the power in the air, or to reduce electrical phenomena to the uses of man;
>
> To build forms, and create machines to contain and distribute the electrical forces of the atmosphere;
>
> To harness the activity of matter, and to drive it towards certain ends;
>
> To employ the electrical force in the air to vitalise, rebuild, and heal the physical body;

then the phenomena of the Rays, working in cycles, will be comprehended, and vast opportunities will be seized by man to bring about specific ends during specific cycles.

A Treatise on Cosmic Fire, pp. 426/7

VI. THE SEVEN PLANES AND THE SEVEN RAYS
The Relationship of the seven planes and the seven rays

- THE EFFECTS OF ROTARY MOTION

Every sphere in the body macrocosmic rotates. This rotation produces certain effects, which effects might be enumerated as follows:

1. *Separation* is produced by rotary movement. By means of this action, all the spheres became differentiated, and form, as we know, the following atomic units:

   a. The solar system, recognised as a cosmic atom, all the so-called atoms within its periphery being regarded as molecular.
   b. The seven planes, regarded as seven vast spheres, rotating *latitudinally* within the solar periphery.
   c. The seven rays, regarded as the seven veiling forms of the Spirits, themselves spheroidal bands of colour, rotating *longitudinally*, and forming (in connection with the seven planes) a vast interlacing network. These two sets of spheres (planes and rays) form the totality of the solar system, and produce its form spheroidal.

Let us withdraw our thought at this juncture from the informing Consciousnesses of these three types of spheres, and concentrate our attention upon the realisation that each plane is a vast sphere of matter, actuated by latent heat and progressing or rotating in one particular direction. Each ray of light, no matter of what colour, is likewise a sphere of matter of the utmost tenuity, rotating in a direction opposite to that of the planes. These rays produce by their mutual interaction a radiatory effect upon each other. Thus by the approximation of the latent heat in matter, and the interplay of that heat upon other spheres that totality is produced which we call "fire by friction."

In connection with these two types of spheres we might, by way of illustration and for the sake of clarity, say that:

a. The planes rotate from east to west.
b. The rays rotate from north to south.

THE SEVEN PLANES AND THE SEVEN RAYS

Students should here bear carefully in mind that we are not referring here to points in space; we are simply making this distinction and employing words in order to make an abstruse idea more comprehensible. From the point of view of the totality of the rays and planes there is no north, south, east nor west. But at this point comes a correspondence and a point of real interest, though also of complexity. By means of this very interaction, the work of the four Maharajahs or Lords of Karma, is made possible; the quaternary and all sumtotals of four can be seen as one of the basic combinations of matter, produced by the dual revolutions of planes and rays.

> The seven planes, likewise atoms, rotate on their own axis, and conform to that which is required of all atomic lives.
>
> The seven spheres of any one plane, which we call subplanes, equally correspond to the system; each has its seven revolving wheels or planes that rotate through their own innate ability, due to latent heat—the heat of the matter of which they are formed.
>
> The spheres or atoms of any form whatsoever, from the form logoic, which we have somewhat dealt with, down to the ultimate physical atom and the molecular matter that goes to the construction of the physical body, show similar correspondences and analogies.

All these spheres conform to certain rules, fulfil certain conditions and are characterised by the same fundamental qualifications.

A Treatise on Cosmic Fire, pp. 152/3

1. SCIENCE AND THE PLANES

• *Plane.* As used in occultism, the term denotes the range or extent of some state of consciousness or of the perceptive power of a particular set of senses or the action of a particular force, or the state of matter corresponding to any of the above.

A Treatise on Cosmic Fire, footnote, p. 66

THE SEVEN PLANES AND THE SEVEN RAYS

CHART III

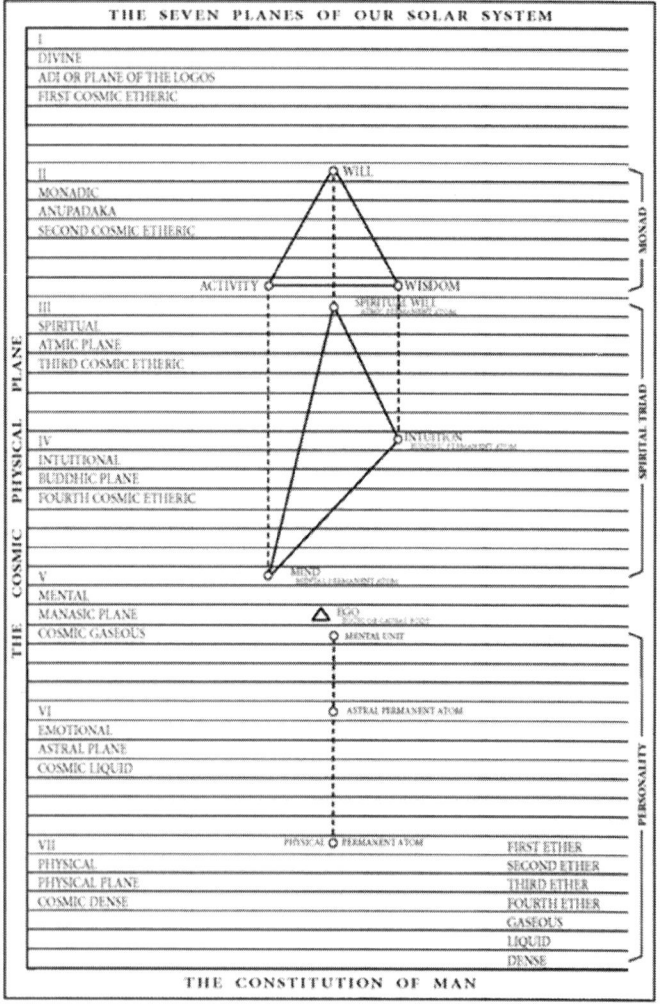

A Treatise on Cosmic Fire, p. 117

THE SEVEN PLANES AND THE SEVEN RAYS

• *I would remind you here* that in *The Secret Doctrine*, H.P.B. states that a plane and a state of consciousness are synonymous terms, and entirely interchangeable; in all my writing I seek to emphasise not the level of matter or substance (a plane, as it is called), but the consciousness which expresses itself in that environing area of conscious substance.   *Esoteric Healing,* pp. 558/9

• Forget not that the seven planes of our solar system constitute the seven subplanes of the cosmic physical plane and that, therefore, spirit is matter at its highest point of expression, and matter is spirit at its lowest. Life differentiates itself into will and love, into great impulsing energies which underlie the entire evolutionary process and motivate its inevitable consummation.   *Esoteric Healing,* p. 589

• We need to bear in mind that our universe (as far as the highest human consciousness can as yet conceive of it) is to be found on the seven subplanes of the cosmic physical plane, and that our highest type of energy, embodying for us the purest expression of Spirit, is but the force manifestation of the first subplane of the cosmic physical plane. We are dealing, therefore, as far as consciousness is concerned, with what might be regarded symbolically as the brain reaction and response to cosmic purpose,—the brain reaction of God Himself.
    Esoteric Psychology II, pp. 5/6

• *Cosmic and Systemic Ethers* - For the sake of those who read this treatise, and because the sequential repetition of fact makes for clarity, let us here briefly tabulate certain fundamental hypotheses that have a definite bearing upon the matter in hand, and which may serve to clear up the present existing confusion concerning the matter of the solar system. Some of the facts stated are already well known, others are inferential, while some are the expression of old and true correspondences couched in a more modern form.

a. The lowest cosmic plane is the cosmic physical, and it is the only one which the finite mind of man can in any way comprehend.

b. This cosmic physical plane exists in matter differentiated into seven qualities, groups, grades, or vibrations.

c. These seven differentiations are the seven major planes of our solar system.

For purposes of clarity, we might here tabulate under the headings physical, systemic, and cosmic, so that the relationship and the correspondences may be apparent, and the connection to that which is above, and to that which is below, or included, may be plainly seen.

THE PLANES

Physical Plane	*Systemic Planes*	*Cosmic Planes*
1. Atomic plane 1st ether	Divine. Adi Primordial matter	Atomic plane 1st ether
2. Sub-atomic	Monadic. Anupadaka The Akasha	Sub-atomic 2nd ether
3. Super-etheric	Spiritual. Atmic Ether	3rd ether

PLANE OF UNION OR AT-ONE-MENT

4. Etheric	Intuitional. Buddhic Air	4th Cosmic ether

THE LOWER THREE WORLDS

5. Gaseous	Mental. Fire	Gaseous -etheric
6. Liquid	Astral. Emotional	Liquid
7. Dense physical	Physical plane	Dense physical

d. These major seven planes of our solar system being but the seven subplanes of the cosmic physical plane, we can consequently see the reason for the emphasis laid by H. P. B. upon the fact that matter and ether are synonymous terms and that this ether is found in some form or other on all the planes, and is but a gradation of cosmic atomic matter, called when undifferentiated mulaprakriti or primordial pregenetic substance, and when differentiated by Fohat (or the energising Life, the third Logos or Brahma) it is termed prakriti, or matter.

e. Our solar system is what is called a system of the fourth order; that is, it has its location on the fourth cosmic etheric plane, counting, as always, from above downwards.

f. Hence this fourth cosmic etheric plane forms the meeting ground for the past and the future, and is the present.

g. Therefore, also, the buddhic or intuitional plane (the correspondence in the system of this fourth cosmic ether) is the meeting ground, or plane of union, for that which is man and for that which will be superman, and links the past with that which is to be.

A Treatise on Cosmic Fire, pp. 116, 118/9

• Religious students will study the side of manifestation we call the "life side" just as the scientist studies that called "matter," and both will come to a realisation of the close relation existing between the two, and thus the old gap and the ancient warfare between science and religion will be in temporary abeyance. Definite methods of demonstrating the fact that life persists after the death of the physical body will be followed, and the etheric web will be recognised as a factor in the case. The connection between the different planes will be sought, and the analogy between the fourth etheric subplane and the fourth or buddhic plane (the fourth cosmic ether) will be studied, for it will be realised that the life of those Entities, Whom we recognise as the planetary Logoi, pours through our scheme from the fourth cosmic plane, the cosmic buddhic, and thus in a very special sense through all lesser correspondences. The alignment will be as follows:

a. The fourth cosmic plane, the cosmic buddhic.
b. The fourth cosmic ether, the fourth plane of the system, the, buddhic.
c. The fourth etheric subplane of our physical plane.

There is thus a line of least resistance from the planes of the cosmos, producing a special activity in connection with the Heavenly Men, demonstrating on Their Own plane, and sequentially in connection with the units in Their bodies on lower levels. Lines of force, extending from our scheme extra-systemically, will be a recognised fact, and will be interpreted by scientists in terms of electrical phenomena, and by the religious man in terms of life,—the life force of certain Entities.

Philosophic students will endeavour simultaneously to link these two schools of thought, and to demonstrate the factor of the intelligent adaptation of the electrical phenomena which we call matter,—of that energised active material we call substance,—to the life purpose of a cosmic Being. In these three lines of thought, therefore,—scientific, religious, and philosophic,—we have the beginning of the *conscious* building, or construction of the antahkarana of that group which we call the fifth rootrace. *A Treatise on Cosmic Fire,* pp. 429/30

• *The Planes and the Three Fires.* On every plane we have, relatively speaking:

  a. Electric fire demonstrating as the prime condition on the higher three.
  b. Fire by friction as the most significant factor on the lower three.
  c. Solar fire, showing forth as the blaze produced by union on the central plane.

In the solar system this is to be seen in connection with a Heavenly Man on the buddhic plane, where They blaze forth through Their etheric centres. In connection with man on the mental plane, we have a similar condition: the three higher subplanes concern the Spirit aspect in the causal body, the three lower subplanes concern primarily the mental sheath, or fire by friction; the fourth subplane is that on which the force centres of the mental body are to be found. So it is on the physical plane for man—his etheric centres being located in matter of the fourth ether. *A Treatise on Cosmic Fire,* p. 522

• *The Planes and Fiery Energy.* It seems desirable that we should here discuss the analogies on each plane, with the seven sub-planes, reminding the student that we are speaking of the planes as the field of evolution of a solar Logos, and not only as a field for the development of man. In the solar system we have:

First, the three higher planes, which have been rightly called the planes of the three aspects; second, the seventh logoic principle is on the first plane, and we can consider it as the impulse in physical matter which produced His body of objectivity.

On the second plane are found the seven Heavenly Men Who are His principal centres of force. There are others, but we are not here concerned with Them. These latter have achieved a certain specific goal, and are the embodiment of centres which are now quiescent or out of manifestation, the logoic kundalini having turned its attention elsewhere. Under another enumeration they make the ten of the esoteric life, and can also be enumerated as twelve, thus forming the twelve-petalled Lotus, or the heart centre in the Body of the ONE ABOUT WHOM NOUGHT BE SAID.

On the fourth cosmic ether the buddhic plane, are found the etheric centres of the Logos. There are to be found the esoteric planets and the Sun, viewed as the centre of the buddhic principles, and from thence the Logos animates His dense physical manifestation.

Finally, on the three lower planes we have His gaseous, liquid, and dense bodies or sheaths, forming in themselves a unity in one peculiar sense; they are as much a coherent whole as the three higher planes form a unified triple expression of the three persons of the Trinity.

We have a similar analogy in the subplanes of each plane in the system, and this will become ever more apparent as man achieves a greater clarity of vision, and can consciously ascertain for himself the truth about the subjective life.

A Treatise on Cosmic Fire, pp. 518/20

• Students will do well to remember in studying the solar system, the planes, the schemes, man and the atom, that the groupings of the lines or streams of energy during the evolutionary cycles fall naturally into four divisions:

1. 1-3-3
2. 4-3
3. 3-4
4. 3-1-3

Division 1 can be understood under the law of correspondences when the nature of the atomic plane of the solar system, the three cosmic etheric planes, and the three planes of human endeavour are investigated in connection with each other.

Division 2 becomes easier of comprehension when the close relation between the four cosmic etheric planes and the three lower planes is grasped. This can be illuminated by a study of the four physical ethers and the three lower subplanes of our physical plane.

Division 3 finds the clue to its mystery in the constitution of the mental plane, with its three formless levels, and its four levels of form.

Division 4 can be grasped as the student arrives at a comprehension of his own nature as a spiritual triad, an egoic body and a threefold lower man. He can likewise approach the first division in a similar manner, and view himself as a primary force or Monad, a triple secondary force or Ego, and a threefold lower energy, or personality, remembering that we are here dealing only with creative energy and with the Brahma aspect of manifestation as it co-ordinates itself with the Vishnu aspect. *A Treatise on Cosmic Fire*, pp. 923/4

• This differentiation of the subplanes of the systemic planes *into a higher three, a lower three, and a central plane of harmony is only so from the standpoint of electrical phenomena, and not from the standpoint of either pure Spirit, or pure substance, viewed apart from each other. It concerns the mystery of electricity, and the production of light.* The three higher planes concern the central Forces or Lives, the three lower concern the lesser Forces or Lives. We must bear this carefully in mind, remembering that to the occultist there is no such thing as substance, but only Force in varying degrees, only Energy of differentiated quality, only Lives emanating from different sources, each distinctive and apart, and only Consciousness producing intelligent effect through the medium of space.

A Treatise on Cosmic Fire, p. 521

• [Kurukshetra]: On the fourth subplane comes a primary blending of the three fiery Lives, producing archetypally that force manifestation of electricity which eventually causes the blazing forth of the Sons of Light on the next plane. In this electrical connotation we have the three higher planes ever embodying the threefold Spirit aspect, the lower three embodying the threefold substance aspect, and then a plane of at-one-ment whereon an approximation is made which, on the path of return, marks the moment of achievement, and the point of triumph. This is succeeded by obscuration. Hence on every plane in the solar system we have a fourth plane whereon the struggle for perfect illumination, and subsequent liberation takes place, the battle ground, the Kurukshetra. Though for man the fourth plane, the buddhic, is the place of triumph, and the goal of his endeavour, for the Heavenly Man it is the battle-ground, while for the solar Logos it is the burning-ground.

A Treatise on Cosmic Fire, pp. 520/1

• *Man*, the lowest type of *coherent* consciousness (using the word "consciousness" in its true connotation as the "One who knows") is but a cell, a minute atom within a group.

A Heavenly Man represents a coherent conscious group.

A solar Logos on His Own plane holds an analogous place to that of a Heavenly Man in a solar system, and from a still higher standpoint to that of a man within the solar system. When the place of the solar planes within the cosmic scheme is duly apprehended it will be recognised that on cosmic levels of a high order the solar Logos is an Intelligence as relatively low in the order of cosmic consciousness as man is in relation to solar consciousness. He is but a cell in the body of the ONE ABOUT WHOM NOUGHT BE SAID. His work parallels on cosmic levels the work of man on the solar planes. He has to undergo on the three lower cosmic planes a process of developing an apprehension of His environment of the same nature as man in the three worlds. This fact should be remembered by all students of this central division of our subject; above all the analogy between the cosmic physical planes and the solar physical planes must be pondered upon. It holds hid the fourfold mystery:

1. The mystery of the Akasha.

2. The secret of the fifth round.

3. The esoteric significance of Saturn, the third planet.

4. The occult nature of cosmic kundalini, or the electrical force of the system.

A Treatise on Cosmic Fire, pp. 295/6

• The etheric body of man holds hid the secret of his objectivity. It has its correspondence on the archetypal plane,—the plane we call that of the divine manifestation, the first plane of our solar system, the plane Adi. The matter of that highest plane is called often the "sea of fire" and it is the root of the akasha, the term applied to the substance of the second plane of manifestation.

A Treatise on Cosmic Fire, p. 79

• Here again is demonstrated the truth of the teaching given by H.P.B..

Electric Fire .... Positive ... Spirit.
Fire by Friction . Negative .. Matter.
Solar Fire ...... Light ..... The two blended and thus producing the objective blaze.

We have thus considered the question of the electrical origin of all manifestation in connection with the four higher subplanes of the solar system—*those four planes which are the four cosmic ethers, and therefore form the body of objectivity of a Heavenly Man in exactly the same sense as the four physical ethers of the solar system form the etheric body of a man.* I have here repeated the fact, as its importance has not yet been grasped by the average occult student; this fact—when conceded and realised—serves in a wonderful way to clarify the whole subject of planetary evolution. . . .

The whole subject of the akasha will be greatly clarified as exoteric science delves into the question of the ethers. As knowledge of the four types of ethers is available, as the vibratory action of these ethers is realised, and as the details concerning their composition, utilisation, light-bearing capacity, and the various angles from which they may be studied become known then paralleling knowledge anent the corresponding four cosmic ethers will be forthcoming. Much concerning them may be deduced from the already apprehended facts which relate to the four solar physical ethers.

For instance, the fourth ether (which is even now being what we might call "discovered"), is at this stage characterised by certain things. I might enumerate a few of these facts with exceeding brevity, as follows:

a. It is the ether which the violet ray uses as a medium.

b. The fourth ether is that whereof the majority of the etheric bodies of men are made.

c. The fourth ether is largely the principal sphere of influence of the "devas of the shadows," or those violet devas which are closely concerned with the physical evolution of man.

d. It is the etheric sphere within which, at a little later date, the human and the deva evolutions will touch.

e. From this fourth etheric sphere the dense physical bodies are created.

f. It is the sphere of physical individualisation. Only when the animal to be individualised was fully conscious on that subplane of the physical plane was it possible to co-ordinate the corresponding spheres on the astral and mental planes and by means of this triple co-ordination to effect the necessary steps which enabled the quaternary to succeed in its effort to approximate the Triad.

g. This fourth ether in this fourth round and on this fourth chain has to be completely mastered and controlled by the Human Hierarchy, the fourth creative. Every unit of the human family has to attain this mastery before the end of this round.

h. It is the sphere wherein the initiations of the threshold are undergone, and the fivefold initiations of the physical plane are entered upon.

Much more might be further added to this list, but I have sought only to point out those which can be easily realised as having a correspondence on the buddhic plane, the fourth cosmic ether. It should be borne in mind that our physical plane in its subplanes has its analogy likewise to the entire cosmic physical plane.

A Treatise on Cosmic Fire, pp. 325/7

• An interesting analogy may here be noted between the fourth cosmic ether, and the fourth ether on the physical plane of the solar system. Both are in process of becoming exoteric—one from the standpoint of man in the three worlds, and the other from the standpoint of a Heavenly Man. The fourth ether is even now being investigated by scientists, and much that they predicate concerning ether, the atom, radium, and the ultimate "protyle" has to do with this fourth ether. It will eventually be brought under scientific formula, and some of its properties, knowledge concerning its range of influence, and its utilisation will become known unto men. Paralleling this, the buddhic plane, the plane of the Christ principle, is gradually becoming known to those advanced beings who are individually able to cognise their place in the body of a Logos of a planetary scheme. The influence of the buddhic plane, and the electrical force that is its peculiar characteristic, are beginning to be felt, and its energy is also beginning to have a definite effect on the egoic bodies of men; the fourth ether of the physical systemic plane is likewise assuming its rightful place in the minds of men, and the electrical force of that subplane is already being adapted and utilised by man in the assistance of the mechanical arts, for methods of transportation, for widespread illumination, and in healing. These four adaptations of electricity:

1. For mechanical uses,
2. For transportation,
3. For illumination,
4. In healing,

are but the working out on the physical plane of paralleling utilisation of buddhic electrical force.

A Treatise on Cosmic Fire, pp. 320/1

2. SCIENCE AND THE RAYS

• The seven rays are the sum total of the divine Consciousness, of the universal Mind; They might be regarded as seven intelligent entities through Whom the plan is working out. They embody divine purpose, express the qualities required for the materialising of that purpose, and They create the forms and are the forms through which the divine idea can be carried forward to completion. Symbolically, They may be regarded as constituting the brain of the divine Heavenly Man . They correspond to the ventricles of the brain, to the seven centres within the brain, to the seven centres of force, and to the seven major glands which determine the quality of the physical body. They are the conscious executors of divine purpose; They are the seven Breaths, animating all forms which have been created by Them to carry out the plan. *Esoteric Psychology* I, pp. 59/60

• . . . it is becoming obvious to the careful student that the emergence of the teaching on the rays has happened at a time when the scientist is announcing the fact that there is naught to be seen and known save energy, and that all forms are composed of energy units and are in themselves expressions of force. *A ray is but a name for a particular force or type of energy, with the emphasis upon the quality which that force exhibits and not upon the force aspect which it creates. This is a true definition of a ray.*

Esoteric Psychology I, pp. 315/6

• The seven rays are therefore embodiments of seven types of force which demonstrate to us the seven qualities of Deity. These seven qualities have consequently a sevenfold effect upon the matter and forms to be found in all parts of the universe, and have also a sevenfold interrelation between themselves.

Esoteric Psychology I, p. 19

• The Seven Logoi embody seven types of differentiated force, and in this Treatise are known under the names of Lords of the Rays. The names of the Rays are

Ray I	Ray of Will or Power . . . . 1st Aspect
Ray II	Ray of Love-Wisdom . . . . 2nd Aspect
Ray III . . .	Ray of Active Intelligence . . . . 3rd Aspect

These are the major Rays.

Ray IV. . . .	Ray of Harmony, Beauty and Art.
Ray V	Ray of Concrete Knowledge or Science.
Ray VI. . . .	Ray of Devotion or of Abstract Idealism.
Ray VII . . .	Ray of Ceremonial Magic or Order..

A Treatise on Cosmic Fire, p. 5

• [Rays 1, 2 and 3 – The Three Rays of Aspect] The three great rays, which constitute the sum total of the divine manifestation, are aspect rays, and this for two reasons:

First, they are, in their totality, the manifested Deity, the *Word* in incarnation. They are the expression of the creative purpose, and the synthesis of life, quality and appearance.

Secondly, they are active in every form in every kingdom, and they determine the broad general characteristics which govern the energy, the quality and the kingdom in question; through them the differentiated forms come into being, the specialised lives express themselves, and the diversity of divine agents fulfill their destiny in the plane of existence allocated to them.

Along these three streams of qualified life-force the creative agencies of God make their presence powerfully felt, and through their activity every form is imbued with that inner evolutionary attribute which must eventually sweep it into line with divine purpose, inevitably produce that type of consciousness which will enable the phenomenal unit to react to its surroundings and thus fulfill its destiny as a corporate part of the whole. Thus intrinsic quality and specific type radiation become possible. The interplay of these three rays determines the outer phenomenal appearance, attracts the unity of life into one or other of the kingdoms in nature, and into one or other of the myriad divisions within that kingdom; the selective and discriminating process is repeated until we have the many ramifications within the four kingdoms, the divisions, groups within a division, families and branches. Thus the creative process, in its wondrous beauty, sequence and unfoldment, stands forth to our awakening consciousness, and we are left awestruck and bewildered at the creative facility of the Great Architect of the Universe.

Looking at all this beauty from a symbolic angle, and thereby simplifying the concept (which is ever the work of the worker in symbols), we might say that Ray I embodies the dynamic idea of God, and thus the Most High starts the work of creation.

Ray II is occupied with the first formulations of the plan upon which the form must be constructed and the idea materialised, and (through the agencies of this great second emanation) the blue prints come into being with their mathematical accuracy, their structural unity and their geometrical perfection. The Grand Geometrician comes thus to the forefront and makes the work of the Builders possible. Upon figure and form, number and sequences will the Temple be built, and so embrace and express the glory of the Lord. The second ray is the ray of the Master Builder.

Ray III constitutes the aggregate of the active building forces, and the Great Architect, with His Builders, organises the material, starts the work of construction, and eventually (as the evolutionary cycle proceeds upon its way) materialises the idea and purpose of God the Father, under the guidance of God the Son. Yet these three are as much a unity as is a human being who conceives an idea, uses his mind and brain to bring his idea into manifestation, and employs his hands and all his natural forces to perfect his concept. The division of aspects and forces is unreal, except for the purpose of intelligent understanding.

The readers of this treatise who really want to profit by this teaching must train themselves ever to think in terms of the whole. The arbitrary tabulations, the divisions into triplicities and septenates, and the diversified enumeration of forces which are seen as emanating from the seven constellations, the ten planets, and the twelve mansions of the zodiac, are but intended to give the student an idea of a world of energies in which he has to play his part.

Esoteric Psychology I, pp. 158/60

• Ten Basic Propositions – In concluding this section of our treatise, and before starting on our real study of the rays, I seek to formulate for you the fundamental propositions upon which all this teaching is founded. They are for me, a humble worker in the Hierarchy, as they are for the Great White Lodge as a whole, a statement of fact and of truth. For students and seekers they must be accepted as an hypothesis:

One: There is one Life, which expresses Itself primarily through seven basic qualities or aspects, and secondarily through the myriad diversity of forms.

Two: These seven radiant qualities are the seven Rays, the seven Lives, who give Their life to the forms, and give the form world its meaning, its laws, and its urge to evolution.

Three: Life, quality and appearance, or spirit, soul and body constitute all that exists. They are existence itself, with its capacity for growth, for activity, for manifestation of beauty, and for full conformity to the Plan. This Plan is rooted in the consciousness of the seven ray Lives.

Four: These seven Lives, Whose nature is consciousness and Whose expression is sentiency and specific quality, produce cyclically the manifested world; They work together in the closest union and harmony, and cooperate intelligently with the Plan of which They are the custodians. They are the seven Builders, Who produce the radiant temple of the Lord, under the guidance of the Mind of the Great Architect of the Universe.

Five: Each ray Life is predominantly expressing Itself through one of the seven sacred planets, but the life of all the seven flows through every planet, including the Earth, and thus qualifies every form. On each planet is a small replica of the general scheme, and every planet conforms to the intent and purpose of the whole.

Six: Humanity, with which this treatise deals, is an expression of the life of God, and every human being has come forth along one line or other of the seven ray forces. The nature of his soul is qualified or determined by the ray Life which breathed him forth, and his form nature is coloured by the ray Life which—in its cyclic appearance on the physical plane at any particular time—sets the quality of the race life and of the forms in the kingdoms of nature. The soul nature or quality remains the same throughout a world period; its form life and nature change from life to life, according to its cyclic need and the environing group condition. This latter is determined by the ray or rays in incarnation at the time.

Seven: The Monad is the Life, lived in unison with the seven ray Lives. One Monad, seven rays and myriads of forms,—this is the structure behind the manifested worlds.

Eight: The Laws which govern the emergence of the quality or soul, through the medium of forms, are simply the mental purpose and life direction of the ray Lords, Whose purpose is immutable, Whose vision is perfect, and Whose justice is supreme.

Nine: The mode or method of development for humanity is self-expression and self-realisation. When this process is consummated the self expressed is the One Self or the ray Life, and the realisation achieved is the revelation of God as the quality of the manifested world and as the Life behind appearance and quality. The seven ray Lives, or the seven soul types, are seen as the expression of one Life, and diversity is lost in the vision of the One and in identification with the One.

Ten: The method employed to bring about this realisation is experience, beginning with individualisation and ending with initiation, thus producing the perfect blending and expression of life-quality-appearance.

This is a brief statement of the Plan. Of this the Hierarchy of Masters in Its seven divisions (the correspondences of the seven rays) is the custodian, and with Them lies the responsibility in any century of carrying out the next stage of that Plan.

Esoteric Psychology I, pp. 141/3

• The following tabulation is an attempt to define that which it is almost impossible to make intelligible in words. From the angle of the illumined occultist it is meaningless, even more than it is to the average student, because as yet the mystery of electricity and the true nature of electrical phenomena (than which there is naught else) is at this time an unrevealed secret, even to the most advanced of the modern scientists.

Ray	Energy	Technique	Quality	Source
1. Power or Will	Grasping		Dynamic Purpose	Dynamically electrified forms.
2. Love-Wisdom	Attracting		Love	Magnetically electrified forms.
3. Intelligent Activity	Selecting		Intellect	Diffusively electrified forms.
4. Beauty or Art	At-one-ing		Unification	Harmonising electrified forms.
5. Science		Differentiating	Discrimination	Crystallising electrified forms.
6. Idealism		Responding	Sensitivity	Fluidic electrified forms.
7. Organisation		Coordinating	Appearance	Physical electrified forms.

That there is such a thing as electricity, that it probably accounts for all that can be seen, sensed and known, and that the entire universe is a manifestation of electrical power,—all this may be stated and is, today, coming to be recognised. But when that has been said, the mystery remains, and will not be revealed, even in partial measure, until the middle of the next century. Then revelation may be possible, as there will be more initiates in the world, and inner vision and inner hearing will be more generally recognised and present. When man arrives at a better understanding of the etheric body and its seven force centres (which are all related to the seven rays, and in their expression show the seven characteristics and techniques which are here tabulated anent the rays) then some further light can intelligibly be thrown upon the nature of the seven types of electrical phenomena which we call the seven rays. *Esoteric Psychology* II, p. 82-3

- *The Principle of Mutation.* Until fourth-dimensional sight is ours, it will scarcely be possible for us to do more than hint at, and get a passing vision of, the complexity and the interweaving in the system. It is not easy for us to do more than grip as a mental concept the fact that the rays, schemes, planets, chains, rounds, races and laws form a unit; seen from the angle of human vision the confusion seems unimaginable, and the key of its solution to be so hidden as to be useless; yet, seen from the angle of logoic sight, the whole moves in unison, and is geometrically accurate. In order to give some idea of the complexity of the arrangement, I would like here to point out that the Rays themselves circulate, the Law of Karma controlling the interweaving. For instance, Ray I may pass around a scheme (if it is the paramount Ray of the scheme) with its first subray manifesting in a chain, its second in a round, its third in a world period, its fourth in a root race, its fifth in a subrace, and its sixth in a branch race. I give this in illustration, and not as the statement of a fact in present manifestation. This gives us some idea of the vastness of the process, and of its wonderful beauty. It is impossible for us, sweeping through on some one Ray, to visualise or in any way to apprehend this beauty; yet, to those on higher levels and with a wider range of vision, the gorgeousness of the design is apparent.

This complexity is for us very much increased because we do not yet understand the principle governing this mutation. Nor is it possible for even the highest human mind in the three worlds to do more than sense and approximate that principle. By mutation I mean the fact that there is a constant changing and shifting, an endless interweaving and interlocking, and a ceaseless ebb and flow, in the dramatic interplay of the forces that stand for the dual synthesis of Spirit and matter. There is constant rotation in the Rays and planes, in their relative importance from the standpoint of time which is the standpoint most closely associated with us. But we can rest assured that there is some fundamental principle directing all the activities of the Logos in His system, and by wrestling to discover the basic principle on which our microcosmic lives rest, we may discover aspects of this inherent logoic principle. This opens to our consideration a wide range of vision, and though it emphasises the complexity of the subject, it also demonstrates the divine magnitude of the scheme, with its magnificent intricacies. *A Treatise on Cosmic Fire,* pp. 597/9

• The coming into power of this fourth Ray at any time (and such an advent may be looked for towards the close of this lesser cycle, which ended in 1924) will produce a corresponding activity in connection with the fourth subplane in each plane, beginning with the fourth physical ether; this will result in the following effects:

First, physical plane scientists will be able to speak with authority anent the fourth ether, even though they may not recognise it as the lowest of the four etheric grades of substance: its sphere of influence and its utilisation will be comprehended, and "force" as a factor in matter, or the electrical manifestation of energy within definite limits, will be as well understood as is hydrogen at this time. Indications of this can already be seen in the discovery of radium, and the study of radioactive substances and of electronic demonstration. This knowledge will revolutionise the life of man; it will put into his hands that which occultists call "power of the fourth order" (on the physical plane). It will enable him to utilise electrical energy for the regulation of his everyday life in a way as yet incomprehensible; it will produce new methods of illuminating, and of heating the world at a small cost and with practically no initial outlay. The *fact* of the existence of the etheric body will be established, and the healing of the dense physical body, via the etheric body, by *force* utilisation and solar radiation, will take the place of the present methods. Healing will then fall practically into two departments:

1. Vitalisation, by means of:
    a. Electricity.
    b. Solar and planetary radiation.
2. Definite curative processes, through the occult knowledge of:
    a. The force centres.
    b. The work of the devas of the fourth ether.

Transportation on sea and land will be largely superseded by the utilisation of air routes and the transit of large bodies through the air, by means of the instantaneous use of the force or energy inherent in the ether itself, will take the place of the present methods.

A Treatise on Cosmic Fire, pp. 428/9

- The sixth ray influence produced the emergence in men's minds of the following knowledges:

1. Knowledge of physical plane light and electricity.
2. Among the esotericists and spiritualists of the world, knowledge of the existence of the astral light.
3. An interest in illumination, both physical and mental.
4. Astro-physics and the newer astronomical discoveries.

The seventh ray will change the theories of the advanced thinkers of the race into the facts of the future educational systems. Education and the growth of the understanding of illumination in all fields will eventually be regarded as synonymous ideals. *Esoteric Psychology* I, pp. 360/1

- The will of Deity coloured the stream of energy units which we call by the name of the Ray of Will or Power, the first ray, and the impact of that stream on the matter of space insured that the hidden purpose of Deity would inevitably and eventually be revealed. It is a ray of such dynamic intensity that we call it the ray of the Destroyer. It is not as yet functioning actively. It will come into full play only when the time comes for the purpose to be safely revealed. Its units of energy in manifestation in the human kingdom are very few. As I earlier said, there is not a true first ray type in incarnation as yet. Its main potency is to be found in the mineral kingdom, and the key to the mystery of the first ray is to be found in radium.

Esoteric Psychology I, p. 44

- . . . I referred earlier to the work of the seventh ray in connection with the phenomena of electricity, through which the solar system is coordinated and vitalised. There is an aspect of electrical phenomena which produces cohesion, just as there is an aspect which produces light. This has not yet been recognised. It is stated in *The Secret Doctrine* of H.P.B. and in *A Treatise on Cosmic Fire*, that the electricity of the solar system is threefold: there is fire by friction, solar fire, and electric fire—the fire of body, of soul and of spirit. Fire by friction is coming to be somewhat understood by the scientists of the world, and we are harnessing to our needs the fire which heats, which gives light, and which produces motion. This is in the physical sense of the words. One of the imminent discoveries will be the integrating power of electricity as it produces the cohesion within all forms and sustains all form life during the cycle of manifested existence. It produces also the coming together of atoms and of the organisms within forms, so constructing that which is needed to express the life principle. Men today are investigating such matters as electro-therapeutics and studying the theory of the electrical nature of the human being. They are working rapidly towards this coming discovery, and much will be revealed along these lines during the next fifty years. The principle of coordination about which men talk has reference, in the last analysis, to this concept, and the scientific basis of all meditation work is really to be found in this basic truth. The bringing in of force and the offering of a channel are all mystical ways of expressing a natural phenomenon as yet little understood, but which will eventually give the clue to the second aspect of electricity. This will be released in fuller measure during the Aquarian age, through the agency of the seventh ray. One of its earliest effects will be the increase of the understanding of brotherhood and its really scientific basis.

Esoteric Psychology I, pp. 373/4

a. *Cosmic Rays*

• The prime cosmic function of the seventh ray is to perform the magical work of blending spirit and matter in order to produce the manifested form through which the life will reveal the glory of God. Students would be well advised to pause here and re-read the section of this treatise in which I dealt with the seventh ray Lord, with His names, and with His purpose. When this has been done, it will be apparent that one of the results of the intensified new influence will be the recognition, by science, of certain effects and characteristics of the work being accomplished. This can already be seen in the work done by scientists in connection with the mineral world. As we have seen in an earlier part of this book the mineral kingdom is governed by the seventh ray, and to the potency of this incoming ray can be attributed the discovery of the radio-activity of matter. The seventh ray expresses itself in the mineral kingdom through the production of radiation, and we shall find that increasingly these radiations (many of which still remain to be discovered) will be noted, their effects understood and their potencies grasped. One point remains as yet unrealised by science, and that is that these radiations are cyclic in their appearance; under the influence of the seventh ray it has been possible for man to discover and work with radium. Radium has always been present, but not always active in such a manner that we were able to detect it. It is under the influence of the incoming seventh ray that its appearance has been made possible, and it is through this same influence that we shall discover new cosmic rays. They too are always present in our universe, but they use the substance of the incoming ray energy as the path along which they can travel to our planet and thus be revealed. It is many thousands of years since what are now studied as the Cosmic Rays (discovered by Millikan) played definitely upon our planet, and at that time the fifth ray was not active as it now is. Therefore scientific knowledge of their activity was not possible.

Other cosmic rays will play upon our earth as this seventh ray activity becomes increasingly active, and the result of their influence will be to facilitate the emergence of the new racial types, and above all else, to destroy the veil or web which separates the world of the seen and tangible from the world of the unseen and the intangible, the astral world. Just as there is a veil called "the etheric web" dividing off the various force centres in the human body, and protecting the head centres from the astral world, so there is a separating web between the world of physical life and the astral world. This will be destroyed, slowly and certainly, by the play of the cosmic rays upon our planet.

The etheric web which is found between the centres in the spine, and which is found at the top of the head (protecting the head centre) is destroyed in man's mechanism by the activity of certain forces found in that mysterious fire which we call the kundalini fire. The cosmic rays of which the modern scientist is aware constitute aspects of the planetary kundalini, and their effect will be the same in the body of the planetary Logos, the Earth, as it is in the human body; the etheric web between the physical and astral planes is in process of destruction, and it is of this event which the sensitives of the world and the spiritualists prophesy as an imminent happening.

Esoteric Psychology I, pp. 369/70

• [Seventh Ray]: People today have no idea what effect this radiation (due to the incoming ray) will have, not only upon the surrounding mineral world but on the vegetable kingdom (which has its roots in the mineral kingdom), and upon men and animals in lesser degree. The power of the incoming cosmic rays has called forth the more easily recognised radio-activity with which modern science is now concerned. It was three seventh ray disciples who "interpreted" these rays to man. I refer to the Curies and to Millikan. Being themselves on the seventh ray, they had the necessary psychic equipment and responsiveness to enable them intuitively to recognise their own ray vibration in the mineral kingdom. *Esoteric Psychology* I, p. 226

VII. COSMIC AND SYSTEMIC LAWS

- *The Laws of Thought.*

There are three great laws, that we might term the fundamental laws of the cosmos, of that greater system (recognised by all astronomers), of which we form a part, and seven laws inherent in the solar system. These seven we might consider secondary laws, though, from the standpoint of humanity, they appear as major ones.

a. Three Cosmic Laws. The first of the cosmic laws is the *Law of Synthesis.* It is almost impossible for those of us who have not the buddhic faculty in any way developed, to comprehend the scope of this law. It is the law that demonstrates the fact that all things—abstract and concrete—exist as one; it is the law governing the thought form of that One of the cosmic Logoi in Whose consciousness both our system, and our greater centre, have a part. It is a unit of His thought, a thought form in its entirety, a concrete whole, and not the differentiated process that we feel our evolving system to be. It is the sumtotal, the centre and the periphery, and the circle of manifestation regarded as a unit.

The second law is the *Law of Attraction and Repulsion*. Fundamentally, the law describes the compelling force of attraction that holds our solar system to the Sirian; that holds our planets revolving around our central unit, the sun; that holds the lesser systems of atomic and molecular matter circulating around a centre in the planet; and that holds the matter of all physical plane bodies, and that of the subtle bodies coordinated around their microcosmic centre.

The third law is the *Law of Economy,* and is the law which adjusts all that concerns the material and spiritual evolution of the cosmos to the best possible advantage and with the least expenditure of force. It makes perfect each atom of time, and each eternal period, and carries all onward, and upward, and *through*, with the least possible effort, with the proper adjustment of equilibrium, and with the necessary rate of rhythm. Unevenness of rhythm is really an illusion of time, and does not exist in the cosmic centre. We need to ponder on this, for it holds the secret of peace, and we need to grasp the significance of that word *through,* for it describes the next racial expansion of consciousness, and has an occult meaning.

In the nomenclature of these laws much is lost, for it is well nigh impossible to resolve abstractions into the terms of speech, and not lose the inner sense in the process. In these laws we again have the threefold idea demonstrated, and the correspondence, as might be expected, holds good.

The Law of Synthesis. . The Will Aspect. . . . . 1st Aspect.
The Law of Attraction. .The Love Aspect. . . . .2nd Aspect.
The Law of Economy. . The Activity Aspect. . 3rd Aspect.

b. Seven Systemic Laws.—Subsidiary to the three major laws, we find the seven laws of our solar system. Again we find the law of analogy elucidating, and the three becoming the seven as elsewhere in the logoic scheme. In each of these seven laws we find an interesting correlation with the seven planes. They are:

1. *The Law of Vibration*, the basis of manifestation, starting on the first plane. This is the atomic law of the system, in the same sense that on each of our planes the first subplane is the atomic plane.
2. *The Law of Cohesion.* On the second plane cohesion is first apparent. It is the first molecular plane of the system, and is the home of the Monad. Divine coherency is demonstrated.
3. *The Law of Disintegration.* On the third plane comes the final casting-off, the ultimate shedding of the sheaths, of the fivefold superman. A Chohan of the sixth Initiation discards all the sheaths beneath the monadic vehicle, from the atmic to the physical.
4. *The Law of Magnetic Control* holds sway paramountly on the buddhic plane, and in the development of the control of this law lies hid the control of the personality by the Monad via the egoic body.
5. *The Law of Fixation* demonstrates principally on the mental plane and has a close connection with manas, the fifth principle. The mind controls and stabilises, and coherency is the result.

6. *The Law of Love* is the law of the astral plane. It aims at the transmutation of the desire nature, and links it up with the greater magnetism of the love aspect on the buddhic plane.

7. *The Law of Sacrifice and Death* is the controlling factor on the physical plane. The destruction of the form, in order that the evolving life may progress, is one of the fundamental methods in evolution.

The Intermediate Law of Karma.—There is also an intermediate law, which is the synthetic law of the system of Sirius. This law is called by the generic term, the Law of Karma, and really predicates the effect the Sirian system has on our solar system. Each of the two systems, as regards its internal economy, is independent in time and space, or (in other words), in manifestation. We have practically no effect on our parent system, the reflex action is so slight as to be negligible, but very definite effects are felt in our system through causes arising in Sirius. These causes, when experienced as effects, are called by us the Law of Karma, and at the beginning they started systemic Karma which, once in effect, constitutes that which is called Karma in our occult and oriental literature.

A Treatise on Cosmic Fire, pp. 567/70

VIII. CREATION – THE BUILDING OF FORM

• The word 'creation' must be occultly understood, and means the appearance in active manifestation of some form of energy.
A Treatise on Cosmic Fire, p. 774

• As time elapses the work of the Heavenly Men in the cosmic etheric spheres will be better comprehended, and assisted intelligently by those lesser intelligences who—by the study of the physical ethers—will eventually hold the key of the greater manifestation. Science is the handmaiden of wisdom, and opens the door to those infinite reaches and to those cosmic expanses, where stand Those vaster Intelligences, Who manipulate the matter of the higher planes, and bend it to the desired form, causing the vibrations thus set up to be felt at the furthest bounds of the solar ring-pass-not. Automatically then all lesser lives and all the denser materials are swept and carried into the needed channels and forms. *Vibration*, or initial activity, *light,* or activity taking form and animating form, *sound* the basis of differentiation and the source of the evolutionary process, and *colour* the sevenfold differentiation—thus is the work carried on. We have been dealing with these four in connection with a solar Logos, and equally with the work of a Heavenly Man and of Man, of the human monad.
A Treatise on Cosmic Fire, p. 329

• The fire latent in matter—itself a product of an earlier manifestation of the same cosmic Identity, or the relatively perfected quality worked out by Him in a previous cosmic incarnation—is set in motion again by the desire of that same Identity to circle once more the wheel of rebirth. That "fire by friction" produces heat and radiation and calls forth a reaction from its opposite "electric fire" or spirit. Here we have the thought of the Ray striking through matter, for the action of electric fire is ever forward, as earlier suggested. The one Ray "electric fire" drops into matter. This is the systemic marriage of the Father and the Mother. The result is the blending of these two

CREATION – THE BUILDING OF FORM

fires, and their united production of that expression of fire which we call "solar fire." Thus is produced the Son. Active Intelligence and Will are united and love-wisdom, when perfected through evolution, will be the outcome. *A Treatise on Cosmic Fire,* pp. 240/1

• In all the constructive work of form-building, certain factors enter in which must here be enumerated as they concern vitally this particular Heavenly Man, and the particular plane, the physical, on which we undergo experience. These are:

First. The will or the one-pointed purpose of some entity.

Second. The material through which the life proposes to manifest. This material, as we know, is found within the ring-pass-not in seven grades, and in forty-nine subgrades.

Third. The Builders who are the vehicle for the divine purpose, and who mould matter upon a particular plan. These Builders evolve the forms out of their own nature and substance.

Fourth. A plan by which the work is carried out and which is imparted to the Builders, being latent in their consciousness. They evolve the form of the Grand Heavenly Man, of the Heavenly Men, of the human units, and of all forms from within outwards, and produce the self-identified Existences as a mother builds and produces a conscious Son out of the matter of her own body, carrying certain racial earmarks yet independent, self-conscious, self-willed and threefold in manifestation. The fact of the identity of the deva evolution with the essence they manipulate must ever be borne in mind.

Finally. Certain Words or Mantric Sounds, which—uttered by a greater Life—can ever drive the lesser lives to the fulfillment of constructive purpose. These Words are uttered by

- A solar Logos. The threefold Word gives rise to a sevenfold vibration.
- A Heavenly Man, Who—through utterance—sweeps into evolutionary objectivity His scheme and all that is therein.
- The Monad, whose threefold word gives rise to a sevenfold vibration.

> The Ego, who—through sonorous utterance—produces a human being in the three worlds.
>
> The analogy existing between these four should be carefully noted. *A Treatise on Cosmic Fire,* pp. 447/9

• What we call the archetypal plane in connection with the Logos (the plane whereon He forms His ideals, His aspirations and His abstract conceptions) is the logoic correspondence to the atomic abstract levels of the mental plane, from whence are initiated the impulses and purposes of the Spirit in man,—those purposes which eventually force him into an objective form, thus paralleling logoic manifestation. First the abstract concept, then the medium provided for manifestation in form, and, finally, that form itself. Such is the process for Gods and for men, and in it is hidden the mystery of mind and of its place in evolution. *A Treatise on Cosmic Fire,* p. 397

• As time goes on, science will become aware of the basic nature and fundamental accuracy of the method whereby every form can be divided into its three aspects, and viewed as an Entity energised by three types of force, emanating from various points extraneous to the form under consideration. It can be considered also as expressing in some way or another, in its various parts, force or energy originating in the three forces of manifestation, Brahma, Vishnu and Shiva. Where this is the case and the premise admitted, the entire outlook on life, on nature, medicine and science and on methods of construction or destruction will be changed. Things will be viewed as essential triplicities, men will be regarded as a combination of energy units, and work with things and with men from the *form* aspect will be revolutionised. *A Treatise on Cosmic Fire*, p. 1186

• . . . it might be wise here to emphasise the necessity of remembering that when we consider the etheric levels of the physical plane we are dealing with those planes upon which the *true form* is to be found, and are approaching the solution of the mystery of the Holy

CREATION – THE BUILDING OF FORM

Spirit and the Mother. In this realisation, and its extension to include an entire solar system, will come a clarifying of the connection between the four higher planes of the system and the three worlds of human endeavour. . . .

The work, therefore, on etheric levels, and the energy and activity originating therefrom, are the factors that primarily are responsible on the physical plane for all that is tangible, objective, and manifested. The accretion of matter around the vital body, and the densification of substance around the vital etheric nucleus are in themselves the result of interaction, and the final interchange of vibration between that which might be called the residue from an earlier manifestation, and the vibration of this present one.

It is here—in the relation between positive electrical energy in its fourfold differentiation, and the triple negative receptive lower substance—that scientists will eventually arrive at certain definite deductions and discover:

a. The secret of matter itself, that is, matter as we know and see it.
b. The key to the process of creation upon the physical plane, and the method whereby density and concretion on the three lower levels are brought about.
c. The formulas for organic transmutation, or the key to the processes whereby the elements as we know them can be disintegrated and recombined.

Only when scientists are prepared to admit the fact that there is a body of vitality which acts as a focal point in every organised form, and only when they are willing to consider each element and form of every degree as constituting part of a still greater vital body, will the true methods of the great goddess Nature become their methods. To do this they must be prepared to accept the sevenfold differentiation of the physical plane as stated by Eastern occultism, to recognise the triple nature of the septenary manifestation.

a. The atomic or Shiva energy, the energy of the first subplane or the first etheric plane.

b. The vital form building energy of the three ensuing etheric levels.
c. The negative receptive energy of the three planes of the dense physical, the gaseous, the liquid and the truly dense.

They will also eventually consider the interplay between the lower three and the higher four in that great atom called the physical plane. This can be seen duplicated in the atom of the physicist or chemist. Scientific students who are interested in these matters will find it worth while to consider the correspondence between these three types of energy, and that which is understood by the words, atoms, electrons, and ions.

All that manifests (from God to man) is the result of these three types of energy or force, of their combination, their interplay, and their psychic action and reaction. During the great cycle of logoic appearance it is the second type of energy which dominates and which is of evolutionary importance, and this is why the etheric body which lies back of all that is visible is the most important. This is equally true of gods, of men and of atoms.

Much time is spent in speculating upon the sources of life, upon the springs of action, and upon the impulses which underlie the creative processes. Hitherto science has worked somewhat blindly and has spent much time investigating the lower three planes. It has dealt principally with the Mother, with the negative receptive matter, and is only now becoming aware of the Holy Spirit aspect, or of the energy which enables that Mother to fulfill her function, and to carry on her work.

Considering the same problem *microcosmically* it may be pointed out that men are only now beginning to be aware of the springs of spiritual action, and of the sources of spiritual life. The energy of the higher planes is only revealing itself as men begin to tread the Way, and to come under the influence of buddhi, which flows from the fourth cosmic etheric plane.

Finally, when scientists are willing to recognise and to co-oper-

CREATION – THE BUILDING OF FORM

ate with the intelligent forces that are to be found on etheric levels, and when they become convinced of the hylozoistic nature of all that exists, their findings and their work will be brought into a more accurate correspondence with things as they really are. This, as has been earlier pointed out, will be brought about as the race develops etheric vision, and the truth of the contentions of the occultist is proved past all controversy. *A Treatise on Cosmic Fire*, pp. 916/9

1. COSMIC FIRE–THE ORIGIN OF ENERGY

• (From The Old Commentary): "The Blessed Ones hide Their threefold nature, but reveal Their triple essence by means of the three great groups of atoms. Three are the atoms, and threefold the radiation. The inner core of fire hides itself and is known only through radiation and that which radiates. Only after the blaze dies out and the heat is no longer felt can the fire be known." *A Treatise on Cosmic Fire,* p. 526

• We purpose in these few introductory remarks to lay down the foundation for a *Treatise on Cosmic Fire*, and to consider the subject of fire both macrocosmically and microcosmically, thus dealing with it from the standpoint of the solar system, and of a human being. This will necessitate some preliminary technicalities which may seem at first perusal to be somewhat abstruse and complicated but which, when meditated upon and studied, may eventually prove illuminating and of an elucidating nature, and which also, when the mind has familiarised itself with some of the details, may come to be regarded as providing a logical hypothesis concerning the nature and origin of energy. We have elsewhere, in an earlier book, touched somewhat upon this matter, but we desire to recapitulate and in so doing to enlarge, thus laying down a broad foundation upon which the subject matter can be built up, and providing a general outline which will serve to show the limits of our discussion. Let us, therefore, look at the subject macrocosmically and then trace the correspondence in the microcosm, or human being.

I. Fire in the Macrocosm

In its essential nature Fire is threefold, but when in manifestation it can be seen as a fivefold demonstration, and be defined as follows:

> 1. *Fire by friction,* or internal vitalising fire. These fires animate and vitalise the objective solar system. They are the sumtotal of logoic kundalini, when in full systemic activity.
> 2. *Solar Fire*, or cosmic mental fire. This is that portion of the cosmic mental plane which goes to the animation of the mental body of the Logos. This fire may be regarded as the sumtotal of the sparks of mind, the fires of the mental bodies and the animating principle of the evolving units of the human race in the three worlds.
> 3. *Electric Fire*, or the logoic Flame Divine. This flame is the distinguishing mark of our Logos, and it is that which differentiates Him from all other Logoi; it is His dominant characteristic, and the sign of His place in cosmic evolution. . . . *A Treatise on Cosmic Fire,* pp. 37/8

• The internal fires that animate and vitalise shew themselves in a twofold manner:

First as l*atent heat.* This is the basis of rotary motion and the cause of the spheroidal coherent manifestation of all existence, from the logoic atom, the solar ring-pass-not, down to the minutest atom of the chemist or physicist.

Second, as *active heat.* This results in the activity and the driving forward of material evolution. On the highest plane the combination of these three factors (active heat, latent heat and the primordial substance which they animate) is known as the 'sea of fire,' of which akasha is the first differentiation of pregenetic matter.

A Treatise on Cosmic Fire, pp. 42/3

CREATION – THE BUILDING OF FORM 65

- *II. Fire in the Microcosm*

Let us briefly consider therefore the correspondence between the greater whole and the unit man . . .

Fire in the Microcosm is likewise threefold in essence and five-fold in manifestation.

1. *There is Internal Vitalising Fire,* which is the correspondence to fire by friction. This is the sumtotal of individual kundalini; it animates the corporeal frame and demonstrates also in the twofold manner:

First, *as latent heat* which is the basis of life of the spheroidal cell, or atom, and of its rotary adjustment to all other cells.

Second, as *active heat* or prana; this animates all, and is the driving force of the evolving form. It shows itself in the four ethers and in the gaseous state, and a correspondence is here found on the physical plane in connection with man to the Akasha and its fivefold manifestation on the plane of the solar system.

This fire is the basic vibration of the little system in which the monad or human spirit is the logos, and it [Page 46] holds the personality or lower material man in objective manifestation thus permitting the spiritual unit to contact the plane of densest matter. It has its correspondence in the ray of intelligent activity and is controlled by the Law of Economy in one of its subdivisions, the Law of Adaptation in Time.

2. There is next the *Fire or Spark of Mind* which is the correspondence in man to solar fire. This constitutes the thinking self-conscious unit or the soul. This fire of mind is governed by the Law of Attraction as is its greater correspondence. Later we can enlarge on this. It is this spark of mind in man, manifesting as spiral cyclic activity, which leads to expansion and to his eventual return to the centre of his system, the Monad—the origin and goal for the reincarnating Jiva or human being. As in the macrocosm this fire also manifests in a twofold manner.

It shows as that intelligent will which links the Monad or spirit with its lowest point of contact, the personality, functioning through a physical vehicle.

It likewise demonstrates, as yet imperfectly, as the vitalising factor in the thought forms fabricated by the thinker. As yet but few thought forms, comparatively, can be said to be constructed by the center of consciousness, the thinker, the Ego. Few people as yet are in such close touch with their higher self, or Ego, that they can build the matter of the mental plane into a form which can be truly said to be an expression of the thoughts, purpose or desire of their Ego, functioning through the physical brain.

A Treatise on Cosmic Fire, pp. 45/6

• *III. Fire in Manifestation.*

To continue our consideration of the fires which sustain the economy of the visible solar system, and of the visible objective human being, which produce evolutionary development, and which are the bases of all objective efflorescence, it must be noted that they demonstrate as the sumtotal of the vital life of a solar system, of a planet, of the entire constitution of active functioning man upon the physical plane, and of the atom of substance.

Speaking broadly we would say that the first fire deals entirely with:

a. Activity of matter.
b. The rotary motion of matter.
c. The development of matter by the means of friction, under the law of Economy. H. P. B. touches on this in the *Secret Doctrine*.

The *second fire,* that from the cosmic mental plane, deals with:

a. The expression of the evolution of mind or manas.
b. The vitality of the soul.
c. The evolutionary expression of the soul as it shows forth in the form of that elusive something which brings about the synthesis of matter. As the two merge by means of this active energising factor, that which is termed *consciousness* appears. As the merging proceeds and the fires become more and more synthesised, that totality of manifestation which we regard as

a conscious existence becomes ever more perfected.
- *d.* The operation of this fire under the Law of Attraction.
- *e.* The subsequent result in the spiral-cyclic movement which we call, within the system, solar evolution, but which (from the standpoint of a cosmos) is the approximation of our system to its central point. This must be considered from the standpoint of time.

The *third fire* deals with:

- *a.* The evolution of spirit.

  Practically nothing can at this stage be communicated anent this evolution. The development of spirit can be only expressed as yet in terms of the evolution of matter, and only through the adequacy of the vehicle, and through the suitability of the sheath, the body or form, can the point of spiritual development reached in any way be appraised. A word of warning should here be interpolated:—

  Just as it is not possible upon the physical plane for the physical vehicle fully to express the total point of development of the Ego or higher self, so it is not possible even for the Ego fully to sense and express the quality of spirit. Hence the utter impossibility for human consciousness justly to appraise the life of the spirit or Monad.

- *b.* The working of the flame divine under the Law of Synthesis—a generic term which will be seen eventually to include the other two laws as subdivisions.
- *c.* The subsequent result of forward progressive motion—a motion which is rotary, cyclic and progressive.

A Treatise on Cosmic Fire, pp. 48/50

2. ETHERIC MATTER

• Through the etheric body, therefore, circulates energy emanating from some mind.

Telepathy and the Etheric Vehicle, p. 3

• It is a fact that omnipresence, which is a law in nature and based on the fact that the etheric bodies of all forms constitute the world etheric body, makes omniscience possible.

Telepathy and the Etheric Vehicle, p. 7

• The etheric body has been described as a network, permeated with fire, or as a web, animated with golden light. It is spoken of in the Bible as the "golden bowl." It is a composition of that matter of the physical plane which we call etheric, and its shape is brought about by the fine interlacing strands of this matter being built by the action of the lesser Builders into the form or mould upon which later the dense physical body can be moulded. Under the Law of Attraction, the denser matter of the physical plane is made to cohere to this vitalised form, and is gradually built up around it, and within it, until the interpenetration is so complete that the two forms make but one unit; the pranic emanations of the etheric body itself play upon the dense physical body in the same manner as the pranic emanations of the sun play upon the etheric body. It is all one vast system of transmission and of interdependence within the system. All receive in order to give, and to pass on to that which is lesser or not so evolved. Upon every plane this process can be seen.

Thus the etheric body forms the archetypal plane in relation to the dense physical body. *A Treatise on Cosmic Fire,* pp. 79/80

• . . .the individual etheric body is not an isolated and separated human vehicle but is, in a peculiar sense, an integral part of the etheric body of that entity which we have called the human family; this kingdom in nature, through its etheric body, is an integral part of the planetary etheric body; the planetary etheric body is not separated off from the etheric bodies of other planets but all of them in their totality, along with the etheric body of the sun constitute the etheric, body of the solar system. This is related to the etheric bodies of the six solar systems which, with ours, form a cosmic unity and into these pour energies and forces from certain great constellations. The field of space is etheric in nature and its vital body is composed of the totality of etheric bodies of all constellations, solar systems and planets which are found therein. Throughout this cosmic golden web there is a constant circulation of energies and forces and this constitutes the scientific basis of the astrological theories. Just as the forces of the planet and of the inner spiritual man (to mention only one factor among many) pour through the etheric body of the individual man upon the physical plane, and condition his outer expression, activities, and qualities, so do the varying forces of the universe pour through every part of the etheric body of that entity we call *space* and condition and determine the outer expression, the activities and qualities of every form found within the cosmic periphery.

Esoteric Astrology, pp. 10/1

• When the Biblical words are used: "In Him we live and move and have our being," we have the statement of a fundamental law in nature and the enunciated basis of the fact which we cover by the rather meaningless word: Omnipresence. Omnipresence has its basis in the substance of the universe, and in what the scientists call the ether; this word "ether" is a generic term covering the ocean of energies which are all inter-related and which constitute that one synthetic energy body of our planet.

. . . it must be carefully borne in mind that the etheric body of every form in nature is an integral part of the substantial form of God Himself—not the dense physical form, but what the esotericists regard as the form-making substance. We use the word God to signify the expression of the One Life which animates every form on the outer objective plane. The etheric or energy body, therefore, of every human being is an integral part of the etheric body of the planet itself and consequently of the solar system. Through this medium, every human being is basically related to every other expression of the Divine Life, minute or great. The function of the etheric body is to receive energy impulses and to be swept into activity by these impulses, or streams of force, emanating from some originating source or other. The etheric body is in reality naught but energy. It is composed of myriads of threads of force or tiny streams of energy, held in relation to the emotional and mental bodies and to the soul by their coordinating effect. These streams of energy, in their turn, have an effect on the physical body and swing it into activity of some kind or another, according to the nature and power of whatever type of energy may be dominating the etheric body at any particular time.

Telepathy and the Etheric Vehicle, p. 2

• *The medium through which the thought currents or impressions* (from no matter what source) *must pass* in order to make an impact upon the human brain *is the planetary etheric body*. This is fundamental in its implications. This etheric vehicle makes all relationships possible, because the individual etheric body is an integral part of the vital body of the planet. This vital body is the medium also of all instinctual reactions, such as an animal will evidence when danger is around. The closer that this etheric body is interwoven (if I may use such a word) with the dense physical vehicle, the clearer will be the instinctual reaction—as in the illustration which I have given and which is based upon millennia of such reactions; the greater also will be the sensitivity and the more aptitude will there be for telepathic contact and recognition of the higher impressions. It might also be added that the etheric body of a disciple or even of an advanced person can be so handled and dealt with that it can reject much that might otherwise impinge upon it, pass through it or use it as a channel.

Telepathy and the Etheric Vehicle, pp. 114/5

• Primarily the functions of the etheric body are three in number:

1. It is the receiver of prana.
2. It is the assimilator of prana.
3. It is the transmitter of prana.

A Treatise on Cosmic Fire, p. 97

• It is with the etheric bodies of all we are dealing, and with their vivification by prana (whether cosmic, solar, planetary or human),

CREATION – THE BUILDING OF FORM

with the organs of reception and with the basis of emanations. Here, therefore, we can arrive at certain dicta anent the etheric body which for purposes of clarity might well be enumerated:

First. The etheric body is the mould of the physical body.

Second. The etheric body is the archetype upon which the dense physical form is built, whether it is the form of a solar system, or of a human body in any one incarnation.

Third. The etheric body is a web or network of fine interlacing channels, formed of matter of the four ethers, and built into a specific form. It forms a focal point for certain radiatory emanations, which vivify, stimulate and produce the rotary action of matter.

Fourth. These pranic emanations when focalised and received, react upon the dense matter which is built upon the etheric scaffolding and framework.

Fifth. This etheric web, during incarnation, forms a barrier between the physical and astral planes, which can only be transcended when consciousness is sufficiently developed to permit of escape. This can be seen in both the microcosm and the macrocosm. When a man has, through meditation and concentration, expanded his consciousness to a certain point he is enabled to include the subtler planes, and to escape beyond the limits of the dividing web. . . .

Sixth. In all the three bodies—human, planetary, and systemic or logoic—will be found a great organ within the organism which acts as the receiver of prana. This organ has its etheric manifestation and its dense physical correspondence. . . .

Seventh. Thus in all the three bodies will the resemblance clearly be seen, and the working out in perfect correspondence is easily demonstrable. . . .

Eighth. When the "will to live" vanishes, then the "Sons of Necessity" cease from objective manifestation. This is logically inevitable, and its working out can be seen in every case of *entified objectivity*. . . . This demonstrates on the physical plane in the withdrawing from out of the top of the head of the radiant etheric body and the consequent disintegration of the physical. The framework goes and the dense physical form falls apart; the pranic life is abstracted bodily from out of the dense sheath, and the stimulation of the fires of matter ceases to be. The latent fire of the atom remains; it is inherent, but the form is made by the action of the two fires of matter—active and latent, radiatory and inherent—aided by the fire of the second Logos, and when they are separated the form falls apart. This is a picture in miniature of the essential duality of all things acted upon by Fohat.

A Treatise on Cosmic Fire, pp. 81/2, 83, 84/5

• As the etheric substratum which is the true substance underlying every tangible form is understood, certain great revolutions will be brought about in the domains of science, of medicine and of chemistry. The study of medicine, for instance, will eventually be taken up from a new angle, and its practice will be built upon a comprehension of the laws of radiation, of magnetic currents, and of the force centres found in men's bodies and their relationship to the force centres and currents of the solar system.

A Treatise on Cosmic Fire, p. xiii

• The whole subject is very involved and difficult but light will begin to dawn before long when science recognises the nature, place and responsibility of the etheric body in man, or of his vital body, and its position as the true form, and basic unit of the dense physical body. When this has been admitted, and the illuminating facts recorded and known, when the connection between the two is grasped, and the necessary deductions and correlations are made, the whole subject of logoic manifestation and the work of the Lives on the four higher planes, with their effect upon the Logoic dense physical plane (our three worlds of expression) will take on a new colouring. The thoughts of men will be revolutionised on the subject of creation; the terms and expressions now used will be corrected, and all will be expressed in

CREATION – THE BUILDING OF FORM

terms of form-building energy, and the three modes of electrical phenomena. This realisation is rapidly on its way but only the generation after the one which the children of the present age are expressing, will see it demonstrated to such an extent as to place etheric electrical phenomena beyond all dispute. This will be done by the coming in of egos who are fully conscious on the etheric levels and who can see all that which is now the subject of speculation. They—from their great numbers and high stage of intellectuality—will rescue the whole matter from the realm of controversy and demonstrate the facts.

A Treatise on Cosmic Fire, pp. 1209/10

• The manifestation of the etheric body in time and space has in it what has been esoterically called "two moments of brilliance." These are, first, the moment prior to physical incarnation, when the descending light (carrying life) is focused in all its intensity around the physical body and sets up a rapport with the innate light of matter itself, to be found in every atom of substance. This focusing light will be found to concentrate itself in seven areas of its ring-pass-not, thus creating seven major centres which will control its expression and its existence upon the outer plane, esoterically speaking. This is a moment of great radiance; it is almost as if a point of pulsating light burst into flame, and as if within that flame seven points of intensified light took shape. This is a high point in the experience of taking incarnation, and precedes physical birth by a very short period of time. It is that which brings on the birth hour. The next phase of the process, as seen by the clairvoyant, is the stage of interpenetration, during which "the seven become the twenty-one and then the many"; the light substance, the energy aspect of the soul, begins to permeate the physical body, and the creative work of the etheric or vital body is completed. The first recognition of this upon the physical plane is the "sound" uttered by the newborn infant. It climaxes the process. The act of creation by the soul is now complete; a new light shines forth in a dark place.

The second moment of brilliance comes in reverse of this process and heralds the period of restitution and the final abstraction of its own intrinsic energy by the soul. The prison house of the flesh is dissolved by the withdrawing of the light and life. The forty-nine fires within the physical organism die down; their heat and light are absorbed into the twenty-one minor points of light; these, in their turn, are absorbed by the major seven centres of energy. Then the "Word of Return" is uttered, and the consciousness aspect, the quality nature, the light and energy of the incarnating man, are withdrawn into the etheric body. The life principle withdraws, likewise, from the heart. There follows a brilliant flaring-up of pure electric light, and the "body of light" finally breaks all contact with the dense physical vehicle, focusses for a short period in the vital body, and then disappears. The act of restitution is accomplished. This entire process of the focusing of the spiritual elements in the etheric body, with the subsequent abstraction and consequent dissipation of the etheric body, would be greatly hastened by the substitution of cremation for burial.

Esoteric Healing, pp. 469/70

• Science, as we know, is fast reaching the point where it will be forced to admit the fact of the etheric body, because the difficulties of refusing to acknowledge it, will be far more insuperable than an admission of its existence. Scientists admit already the fact of etheric matter; the success of photographic endeavor has demonstrated the reality of that which has hitherto been considered unreal, because (from the standpoint of the physical) intangible. Phenomena are occurring all the time which remain in the domain of the supernatural unless accounted for through the medium of etheric matter, and in their anxiety to prove the spiritualists wrong, scientists have aided the cause of the true and higher spiritism by falling back on reality, and on the fact of the etheric body, even though they consider it a body of emanative radiation—being concerned with the effect and not having yet ascertained the cause. Medical men are beginning to study (blindly as yet) the question of vitality, the effect of solar rays upon the physical organism, and the underlying laws of inherent and radiatory heat.

CREATION – THE BUILDING OF FORM

They are beginning to ascribe to the spleen functions hitherto not recognized, to study the effect of the action of the glands, and their relation to the assimilation of the vital essences by the bodily frame. They are on the right road, and before long (perhaps within this century) the FACT of the etheric body and its basic function will be established past all controversy, and the whole aim of preventive and curative medicine will shift to a higher level.

A Treatise on Cosmic Fire, pp. 88/9
See also: 'Devas of the Etheric Double',
A Treatise on Cosmic Fire, pp. 925/36

a. *Etheric Centres*

• ...within the human etheric body there are to be found seven major force centres which are in the nature of distributing agencies and electrical batteries, providing dynamic force and qualitative energy to the man; they produce definite effects upon his outer physical manifestation. Through their constant activity his quality appears, his ray tendencies begin to emerge and his point in evolution is clearly indicated.

This "control of form through a septenate of energies" (as it is defined in the Old Commentary) is an unalterable rule in the inner government of our universe and of our particular solar system, as well as in the case of individual man. There are, for instance, in our solar system, seven sacred planets which correspond to the seven individual force centres in man, the seven solar systems, of which our solar system is one, and in their turn the seven energy centres of the One to Whom I have referred in my other books as the One About Whom Naught Can Be Said.   *Esoteric Astrology,* pp. 11/2

• These energies galvanise the physical man into activity through the medium of certain force centres in the etheric body. These, for our immediate purposes, can be divided into three centres below the diaphragm and four above.

These are:
I. Below the diaphragm:
1. The base of the spine.
2. The sacral centre.
3. The solar plexus.

II. Above the diaphragm:
1. The heart centre.
2. The throat centre.
3. The centre between the eyebrows.
4. The head centre. *Esoteric Psychology* I, p. 288

• The centres in the human being deal fundamentally with the FIRE aspect in man, or with his divine spirit. They are definitely connected with the Monad, with the will aspect, with immortality, with existence, with the will to live, and with the inherent powers of Spirit. They are not connected with objectivity and manifestation, but with force, or the powers of the divine life. The correspondence in the Macrocosm can be found in the force which manipulates the cosmic nebulae and which by its whirling rotary motion eventually builds them into planets or spheroidal bodies. These planets are each of them an expression of the "will to live" of some cosmic entity, and the force that swirled, that rotated, that built, that solidified, and that continues to hold in form coherent, is the force of some cosmic Being.

This force originates on cosmic mental levels, from certain great foci there, descends to the cosmic astral, forming corresponding cosmic focal points, and on the fourth cosmic etheric level (the buddhic plane of our solar system) finds its outlet in certain great centres. These centres are again reflected or reproduced in the three worlds of human endeavor. The Heavenly Men, therefore, have centres on three solar planes, a fact to be remembered.

a. On the monadic plane, the plane of the seven Rays.
b. On the buddhic plane, where the Masters and their disciples form the forty-nine centres in the bodies of the seven Heavenly Men.

c. On the fourth etheric physical plane, where the sacred planets, the dense bodies in etheric matter of the Heavenly Men, are to be found.

Here again we can trace the microcosmic correspondence: In the human being the centres are found on the mental plane from which originates the impulse for physical plane existence, or the will to incarnate; to the astral level, and eventually to the etheric levels, to the fourth ether, where they practically go through the same evolution that the planetary centres went through, and are instrumental in bringing about objectivity,—being the force centres.

The centres are formed entirely of streams of force, pouring down from the Ego, who transmits it from the Monad. In this we have the secret of the gradual vibratory quickening of the centres as the Ego first comes into control, or activity, and later (after initiation) the Monad, thus bringing about changes and increased vitality within these spheres of fire or of pure life force.

A Treatise on Cosmic Fire, pp. 165/6

3. PRANA

• "*Prana* is the name by which we designate a universal principle, which principle is the essence of all motion, force or energy, whether manifested in gravitation, electricity, the revolution of the planets, and all forms of life, from the highest to the lowest. It may be called the soul of Force and Energy in all their forms, and that principle which, operating in a certain way, causes that form of activity which accompanies life." (Ramacharaka, Yogi, *The Hindu-Yogi Science of Breath,* pp. 16, 17)   *The Soul and its Mechanism*, p. 98

• In the study of the etheric body and prana lies the revelation of the effects of those rays of the sun which (for lack of better expression), we will call "solar pranic emanations." These solar pranic emanations are the produced effect of the central heat of the sun approaching other bodies within the solar system by one of the three main channels of contact, and producing on the bodies then contacted certain effects differing somewhat from those produced by the other emanations. These effects might be considered as definitely stimulating and constructive, and (through their essential quality) as producing conditions that further the growth of cellular matter, and concern its adjustment to environing conditions; they concern likewise the internal health (demonstrating as the heat of the atom and its consequent activity) and the uniform evolution of the form of which that particular atom of matter forms a constituent part. Emanative prana does little in connection with form building; that is not its province, but it conserves the form through the preservation of the health of its component parts. Other rays of the sun act differently, upon the forms and upon their substance. Some perform the work of the Destroyer of forms, and others carry on the work of cohering and of attracting; the work of the Destroyer and of the Preserver is carried on under the Law of Attraction and Repulsion. Some rays definitely produce accelerated motion, others produce retardation. The ones we are dealing with here—pranic solar emanations—work within the four ethers, that matter which (though physical) is not as yet objectively visible to the eye of man. They are the basis of all physical plane life considered solely in connection with the life of the physical plane atoms of matter, their inherent heat and their rotary motion. These emanations are the basis of that "fire by friction" which demonstrates in the activity of matter. *A Treatise on Cosmic Fire,* pp. 78/9

• *The Transmitter of Prana* - We have dwelt on facts which might stimulate interest and emphasise the utility of this pranic vehicle. Certain facts need emphasis and consideration as we study this static ring and its circulating fires. Let me briefly recapitulate for the sake of clarity:

The System receives prana from cosmic sources via three centres, and redistributes it to all parts of its extended influence, or to the bounds of the solar etheric web. This cosmic prana becomes colored by solar quality and reaches the furthest confines of the system. Its mission might be described as the vitalisation of the vehicle which is the physical material expression of the solar Logos.

CREATION – THE BUILDING OF FORM

The Planet receives prana from the solar centre, and redistributes it via the three receiving centres to all parts of its sphere of influence. This solar prana becomes colored by the planetary quality and is absorbed by all evolutions found within the planetary ring-pass-not. Its mission might be described as the vitalisation of the vehicle which is the physical material expression of one or other of the seven Heavenly Men.

The Microcosm receives prana from the sun after it has permeated the planetary etheric vehicle, so that it is solar prana, plus planetary quality. Each planet is the embodiment of some one ray aspect, and its quality is marked predominantly on all its evolution.

Prana, therefore, which is active radiatory heat, varies in vibration and quality according to the receiving Entity. Man passes the prana through his etheric vehicle, colors it with his own peculiar quality, and so transmits it to the lesser lives that make up his little system. Thus, the great interaction goes on, and all parts blend, merge and are interdependent; and all parts receive, color, qualify and transmit. An endless circulation goes on that has neither a conceivable beginning nor possible end from the point of view of finite man, for its source and end are hid in the unknown cosmic fount. Were conditions everywhere perfected this circulation would proceed unimpeded and might result in a condition of almost endless duration, but limitation and termination result as the effects of imperfection giving place to a gradual perfection. Every cycle originates from another cycle of a relative completeness, and will give place ever to a higher spiral; thus eventuate periods of apparent relative perfection leading to those which are still greater. *A Treatise on Cosmic Fire*, pp. 101/2

- 1. *Solar prana.*

This is that vital and magnetic fluid which radiates from the sun, and which is transmitted to man's etheric body through the agency of certain deva entities of a very high order, and of a golden hue. It is passed through their bodies and emitted as powerful radiations, which are applied direct through certain plexi in the uppermost part of the etheric body, the head and shoulders, and passed down to the etheric correspondence of the physical organ, the spleen, and from thence forcibly transmitted into the spleen itself. These golden hued pranic entities are in the air above us, and are specially active in such parts of the world as California, in those tropical countries where the air is pure and dry, and the rays of the sun are recognised as being specially beneficial. . . .

These solar devas take the radiatory rays of the sun which reach from its centre to the periphery along one of the three channels of approach, pass them through their organism and focalise them there. They act almost as a burning glass acts. These rays are then reflected or transmitted to man's etheric body, and caught up by him and again assimilated. When the etheric body is in good order and functioning correctly, enough of this prana is absorbed to keep the *form organised*. This is the whole object of the etheric body's functioning, and is a point which cannot be sufficiently emphasised. The remainder is cast off in the form of animal radiation, or physical magnetism—all terms expressing the same idea. Man therefore repeats on a lesser scale the work of the great solar devas, and in his turn adds his quota of repolarised or remagnetised emanation to the sumtotal of the planetary aura.

2. *Planetary prana.*

This is the vital fluid emanated from any planet, which constitutes its basic coloring or quality, and is produced by a repetition within the planet of the same process which is undergone in connection with man and solar prana. The planet (the Earth, or any other planet) absorbs solar prana, assimilates what is required, and radiates off that which is not essential to its well-being in the form of planetary radiation. Planetary prana, therefore, is solar prana which has passed throughout the planet, has circulated through the planetary etheric body, has been transmitted to the dense physical planet, and has been cast off thence in the form of a radiation of the same essential character as solar prana, *plus the individual and distinctive quality of the particular planet concerned.* This again repeats the process undergone in the human body. The physical radiations of men differ according to the quality of their physical bodies. So it is with a planet.

Planetary emanative prana (as in the case of solar prana) is caught up and transmitted via a particular group of devas, called the "devas of the shadows," who are ethereal devas of a slightly violet hue. Their bodies are composed of the matter of one or other of the four ethers, and they focalise and concentrate the emanations of the planet, and of all forms upon the planet. They have a specially close connection with human beings owing to the fact of the essential resemblance of their bodily substance to man's etheric substance, and because they transmit to him the magnetism of "Mother Earth" as it is called. Therefore we see that there are two groups of devas working in connection with man:

a. Solar devas, who transmit the vital fluid which circulates in the etheric body.

b. Planetary devas of a violet color, who are allied to man's etheric body, and who transmit earth's prana, or the prana of whichever planet man may be functioning upon during a physical incarnation. . . .

3. *The prana of forms.*

It must first be pointed out that forms are necessarily of two kinds, each having a different place in the scheme:

Forms that are the result of the work of the third and the second Logos, and Their united life. Such forms are the units in the vegetable, animal and mineral kingdoms.

Forms that are the result of the united action of the three Logoi, and comprise the strictly deva and human forms.

There is also the still simpler form embodied in the substance of which all the other forms are made. This matter is strictly speaking the atomic and molecular matter, and is animated by the life or energy of the third Logos.

In dealing with the first group of forms, it must be noted that the pranic emanations given off by units of the animal and vegetable kingdom (after they have absorbed both solar and planetary prana) are naturally a combination of the two, and are transmitted by means of *surface radiation,* as in solar and planetary prana, to certain lesser groups of devas of a not very high order, who have a curious and intricate relationship to the group soul of the radiating animal or vegetable. This matter cannot be dealt with here. These devas are also of a violet hue, but of such a pale color as to be almost grey; they are in a transitional state, and merge with a puzzling confusion with groups of entities that are almost on the involutionary arc.

In dealing with the second group, the human form transmits the emanative radiations to a much higher grade of deva. These devas are of a more pronounced hue, and after due assimilation of the human radiation, they transmit it principally to the animal kingdom, thus demonstrating the close relationship between the two kingdoms. If the above explanation of the intricate inter-relation between the sun and the planets, between the planets and the evolving forms upon them, between the forms themselves in ever descending importance demonstrates nothing more than the exquisite interdependence of all existences, then much will have been achieved.

Another fact which must also be brought out is the close relationship between all these evolutions of nature, from the celestial sun down to the humblest violet *via the deva evolution* which acts as the transmitting transmuting force throughout the system.

Lastly, all work with fire. Fire internal, inherent and latent; fire radiatory and emanative; fire generated, assimilated and radiated; fire vivifying stimulating, and destroying; fire transmitted, reflected, and absorbed; fire, the basis of all life; fire, the essence of all existence; fire, the means of development, and the impulse behind all evolutionary process; fire, the builder, the preserver and the constructor; fire, the originator, the process and the goal; fire the purifier and the consumer. The God of Fire and the fire of God interacting upon each other, till all fires blend and blaze and till all that exists, is passed through the fire—from a solar system to an ant—and emerges as a triple perfection. Fire then passes out from the ring-pass-not as perfected essence, whether essence emerging from the human ring-pass-not, the planetary ring-pass-not or the solar. The wheel of fire turns and all within that wheel is subjected to the threefold flame, and eventually stands perfected.

A Treatise on Cosmic Fire, pp. 90, 91/3, 95/7

CREATION – THE BUILDING OF FORM

4. LIFE

• "The secret of the Fire lies hid in the second letter of the Sacred Word. The mystery of life is concealed within the heart. When the lower point vibrates, when the Sacred Triangle glows, when the point, the middle centre and the apex, connect and circulate the Fire, when the threefold apex likewise burns, then the two triangles—the greater and the lesser—merge into one Flame, which burneth up the whole."
A Treatise on Cosmic Fire, p. 1231

• As to the significance of the word "life" our task is well nigh insuperable, for no human being has, or can have, any comprehension of the nature of life until he has attained the third initiation. I repeat this with emphasis, and in order to impress upon you the futility of idle speculation upon this subject. Disciples who have undergone the third initiation and have climbed the mount of Transfiguration can—from that high point—glimpse the radiance of the subjective centre of energy (the central spiritual sun of *The Secret Doctrine*) and so gain a flash of realisation as to the meaning of the word "life." But they cannot, and they dare not, pass on the knowledge gained. Their efforts to convey such information would be futile, and language itself would be inadequate to the task. Life is not what anyone has hitherto surmised. Energy (in contradistinction to force, and using the word to express the emanating centre which differentiates into forces) is not what idle speculation has portrayed it to be. Life is the synthesis of all activity—an activity which is a blend of many energies, for life is the sum total of the energies of the seven solar systems, of which our solar system is but one. These, in their totality, are the expression of the activity of that Being Who is designated in our hierarchical archives as the "One About Whom Naught May Be Said." This seven-fold cosmic energy, the fused and blended energies of seven solar systems, including ours, sweeps automatically through each of the seven, carrying the qualities of

1. Impulse towards activity.
2. Active impulse towards organisation.
3. Active organised impulse towards a definite purpose.

CREATION – THE BUILDING OF FORM

I have worded these impulses as above in order to show the emergent tendency through their mutual interplay. This triple energetic impulse, borne on the impetus of the seven great breaths or rays, started the world process of Becoming, and manifested as the urge towards evolution,—towards an evolution which is active, organised, and which works undeviatingly and unerringly towards a specific goal. This goal is known in its fullest measure only to that incomprehensible Existence Who works through seven solar systems (in their turn the expression of seven great Lives) just as our solar Deity works through the seven planetary Logoi.

Esoteric Psychology I, pp. 150/1

• It might be stated that the esotericist is occupied in discovering and working with those principles which energise each level of the cosmic physical plane and which are, in reality, aspects of the qualified life energy which is working in and through unprincipled substance. His task is to shift the focus of his attention away from the substance-form side of existence and to become aware of that which has been the source of form production on any specific level. It is his task to develop within himself the needed responsiveness and sensitivity to the quality of the life dominating any form until he arrives eventually at the quality of the ONE LIFE which animates the planet and within Whose activity we live and move and have our being.

To do this, he must first of all discover the nature of his own qualified energies (and here the nature of the governing rays enters in) which are expressing themselves through his three lower vehicles of manifestation, and later through his integrated personality. Having arrived at a measure of this knowledge and having oriented himself towards the qualified life aspect, he begins to develop the subtle, inner mechanism through which contact can be made with the more general and universal aspects. He learns to differentiate between the quality or karmic predispositions of the "unprincipled" substance of which his form and all forms are made, and the qualified principles which are seeking expression through those forms and, incidentally, to redeem, salvage and purify them so that the substance of the next solar system will be of a higher order than that of the present one, and consequently more responsive to the will aspect of the Logos.

Viewed from this angle, *esotericism is the science of redemption,* and of this all World Saviours are the everlasting symbol and exponents. It was to redeem substance and its forms that the planetary Logos came into manifestation, and the entire Hierarchy with its great Leader, the Christ (the present world Symbol), might be regarded as a hierarchy of redeemers, skilled in the science of redemption. Once They have mastered this science, They can then pass on to the Science of Life and deal with the energies which will eventually hold and use the qualified, redeemed and then principled substance and forms. It is the redemption of unprincipled substance, its creative restoration and spiritual integration, which is Their goal; the fruits of Their labour will be seen in the third and final solar system. Their activity will produce a great spiritual and planetary fusion, of which the fusion of personality and soul (at a certain point upon the path of evolution) is the symbol in the microcosmic sense. You can see by this the close relation between the work of the individual aspirant or disciple as he redeems, salvages and purifies his threefold body of manifestation and the work of the planetary Logos as He performs a similar task in connection with the "three periodical vehicles" through which He works: His personality vehicle, His soul expression and His monadic aspect.

By means of all that I have said you will realise that I am endeavouring to take the vagueness out of the word "esotericism," and to indicate the extremely scientific and practical nature of the enterprise upon which all esotericists are embarked.

Education in the New Age, pp. 64/6

• . . . the planet constitutes the response apparatus of a superhuman Life, and that Life responds consciously to impacts emanating from the solar system as a whole, and from certain constellations (embodied Lives) with which our solar system is linked. Similarly the solar Logos functions through the medium of that gigantic response apparatus which is bounded by the ring-pass-not of a solar system. Each form, from that of the tiniest atom to that of a vast constellation, is an embodiment of a life, which expresses itself as consciousness, awareness, and responsive sentiency through the medium of some type of response mechanism. Thus we have the establishing of a universe of lives, interacting and interrelated, all of them conscious, some of them self-conscious, and others group-conscious, but all grounded in the universal mind, all possessing souls, and all presenting aspects of the divine Life. *Esoteric Psychology* I, p.136

• Back of matter, there can be found an immanent and potent factor which is responsible for the coherence of the form nature, and which constitutes the acting personality in the physical world. This can be regarded as the life aspect, and scholars are wrestling all the time with the problem of life, trying to arrive at its origin and its cause.
From Intellect to Intuition, p.178

• Each form (because it constitutes an aggregated area of substantial lives or atoms) is a centre within the etheric body of the form of which it is a constituent part. It has, as the basis of its existence, a living dynamic point which integrates the form and preserves it in essential being. This form or centre—large or small, a man or an atom of substance—is related to all other forms and expressing energies in the environing space, and is automatically receptive to some, and repudiates others through the process of non-recognition; it relays or transmits other energies, radiating from other forms, and it thus becomes in its turn an impressing agent. You see, therefore, where differentiated truths approach each other and blend, forcing us to use the same terminologies in order to express the same factual truths or ideas.

Again, each point of life within a centre has its own sphere of radiation or its own extending field of influence; this field is necessarily dependent upon the type and the nature of the indwelling Consciousness. . . . Bear in mind also that the life which pours through all centres and which animates the whole of space is *the life of an Entity*; it is, therefore, the same life in all forms, limited in time and space by the intention, the wish, the form and the quality of the indwelling

CREATION – THE BUILDING OF FORM

consciousness: the types of consciousness are many and diverse, yet life remains ever the same and indivisible, for it is the ONE LIFE.

The sphere of radiation is conditioned always by the point of evolution of the life within the form; the correlating, integrating factor, relating centre to centre, is life itself; life establishes contact; livingness is the basis of every relation, even if this is not immediately apparent to you; consciousness qualifies the contact and colours the radiation. Thus again we are returned to the same fundamental triplicity to which I gave the names of Life, Quality, Appearance in an earlier book (*Esoteric Psychology* I). A form is therefore a centre of life within some aspect of the etheric body of the Entity, Space, where a living animated existence, such as that of a planet, is concerned. The same is true also of all lesser forms, such as those found upon and within a plane. *Telepathy and the Etheric Vehicle,* pp. 179/82

• H. P. B. says in the *Secret Doctrine:*

"Occultism does not accept anything inorganic in the Cosmos. The expression employed by Science 'inorganic substance' means simply that the latent life, slumbering in the molecules of so-called 'inert matter' is incognisable. All is Life, and every atom of even mineral dust is a Life, though beyond our comprehension and perception... Life therefore is everywhere in the Universe... wherever there is an atom of matter, a particle or a molecule, even in its most gaseous condition, there is life in it however latent and unconscious."

A Treatise on Cosmic Fire, footnote, p. 601

5. EVOLUTION–INVOLUTION

• "Evolution is a continually accelerating march of all the particles of the universe which leads them simultaneously, by a path sown with destruction, but uninterrupted and unpausing, from the material atom to that universal consciousness in which omnipotence and omniscience are realised: in a word, to the full realisation of the Absolute of God."

This proceeds from those minute diversifications which we call molecules and atoms up to their aggregate as they are built into forms;

and continues on through the building of those forms into greater forms, until you have a solar system in its entirety. All has proceeded under law, and the same basic laws govern the evolution of the atom as the evolution of a solar system. The macrocosm repeats itself in man, the microcosm, and the microcosm is again reflected in all lesser atoms. *The Consciousness of the Atom,* pp. 53/4

• It may therefore seem to some of us a logical hypothesis that just as the atom of chemistry is a tiny sphere, or form, with a positive nucleus, which holds rotating around it the negative electrons, so all forms in all the kingdoms of nature are of a similar structure, differing only in degree of consciousness or intelligence. We can therefore regard the kingdoms themselves as the physical expression of some great subjective life, and can by logical steps come to the recognition that every unit in the human family is an atom in the body of that greater unit who has been called in some of the Scriptures the "Heavenly Man." Thus we arrive finally at the concept that the solar system is but the aggregate of all kingdoms and all forms, and the Body of a Being Who is expressing Himself through it, and utilising it in order to work out a definite purpose and central idea. In all these extensions of our final hypothesis, the same triplicity can be seen; an informing Life or Entity manifesting through a form, or a multiplicity of forms, and demonstrating discriminative intelligence.

It is not possible to deal with the method whereby the forms are built up, or to enlarge upon the evolutionary process by means of which atoms are combined into forms, and the forms themselves collected into that greater unity which we call a kingdom in nature. This method might be briefly summed up in three terms—*involution,* or the involving of the subjective life in matter, the method whereby the indwelling Entity takes to itself its vehicle of *expression;* evolution, or the utilisation of the form by the subjective life, its gradual perfecting, and the final liberating of the imprisoned life; and the law of *attraction and repulsion*, whereby matter and spirit are co-ordinated whereby the central life gains experience, expands its consciousness, and, through the use of that particular form attains self-knowledge

CREATION – THE BUILDING OF FORM

and self-control. All is carried forward under this basic law. In every form you have a central life, or idea, coming into manifestation, involving itself more and more in substance, clothing itself in a form and shape adequate to its need and requirement, utilising that form as a means of expression, and then—in due course of time—liberating itself from the environing form in order to acquire one more suited to its need. Thus through every grade of form, spirit or life progresses, until the path of return has been traversed and the point of origin achieved. This is the meaning of evolution and here lies the secret of the cosmic incarnation. Eventually spirit frees itself from form, and attains liberation plus developed psychical quality and graded expansions of consciousness.   *The Consciousness of the Atom,* pp. 61/3

• Again another correspondence between the fourth cosmic ether and the fourth physical ether lies in the fact that they are both primarily concerned with the work of the great builders, bearing in mind that they build the *real* body of the Logos in *etheric* matter; the dense physical vehicle is not so much the result of their work as it is the result of the meeting of the seven streams of force or electricity, which causes that apparent congestion in matter that we call the dense physical planes (the three lower subplanes). This apparent congestion is, after all, but the exceeding electronic activity or energy of the mass of negative atoms awaiting the stimulation that will result from the presence of a certain number of positive atoms. This needs to be borne in mind. The work of evolution is based on two methods and demonstrates as:

Involution, wherein the negative electrons of matter preponderate. The percentage of these feminine electrons is one of the secrets of initiation and is so vast during the involutionary stage that the rarity of the positive atoms is very noticeable; they are so rare as only to serve to keep the mass coherent.

Evolution, wherein, due to the action of manas, these negative atoms become stimulated and either dissipate back into the central electrical reservoir, or merge in their opposite pole, and are consequently again lost. This results in:

Synthesis.

Homogeneity.

The rarity instead of the density of matter. The fourth cosmic ether, the buddhic, is the plane of air, and is also the plane of absorption for the three worlds. This rarefication of dense matter (as we know it) simply means that at the close of the evolutionary process it will have been transmuted and be practically, from our point of view, non-existent; all that will be left will be the positive atoms, or certain vortices of force which—having absorbed the negative will demonstrate as electrical phenomena of a form inconceivable to man at his present stage of knowledge. These vortices will be distinguished by:

1. Intense vibratory activity.
2. The predominance of one certain colour according to the quality of the etheric display, and its source.
3. Repulsion to all bodies of similar vibratory rate and polarity. Their attractive quality at the end of evolution will cease owing to the fact that naught remains to be attracted.

A Treatise on Cosmic Fire, pp. 321/2

6. CONSCIOUSNESS–MIND–MANAS

• . . . the object for which life takes form and the purpose of manifested being is the unfoldment of consciousness, or the revelation of the soul. This might be called the Theory of the Evolution of Light. When it is realised that even the modern scientist is saying that light and matter are synonymous terms, thus echoing the teaching of the East, it becomes apparent that through the interplay of the poles, and through the friction of the pairs of opposites light flashes forth. The goal of evolution is found to be a gradual series of light demonstrations. Veiled and hidden by every form lies light. As evolution proceeds, matter becomes increasingly a better conductor of the light, thus demonstrating the accuracy of the statement of the Christ "I am the Light of the World".   *A Treatise on White Magic,* pp. 9/10

• Love is the great unifier, the prime attractive impulse, cosmic and microcosmic, but the mind is the main creative factor and the utiliser of the energies of the cosmos. Love attracts, but the mind attracts, repels and co-ordinates, so that its potency is inconceivable.
A Treatise on White Magic, p. 125

• An embodied idea, therefore, is literally a positive impulse, emanating from mental levels, and clothing itself in a veil of negative substance. These two factors in turn will be regarded as emanations from a still greater force centre, which is expressing purpose through them both.

A thought form, as constructed by man, is the union of a positive emanation and a negative. These two are the emanations of a Unity, the coherent Thinker. *A Treatise on Cosmic Fire*, p. 560

• Consciousness might be defined as the faculty of apprehension, and concerns primarily the relation of the Self to the not-self, of the Knower to the Known, and of the Thinker to that which is thought about. All these definitions involve the acceptance of the idea of duality, of that which is objective and of that which lies back of objectivity.

Consciousness expresses that which might be regarded as the middle point in manifestation. It does not involve entirely the pole of Spirit. It is produced by the union of the two poles, and the process of interplay and of adaptation that necessarily ensues.
A Treatise on Cosmic Fire, p. 243

• Consciousness is inherent in all forms of life. That is an occult platitude. It is an innate potency which forever accompanies life in manifestation. *Telepathy and the Etheric Vehicle*, p. 65

• DEFINITIONS OF MANAS OR MIND

1. *Manas, as we already know, is the fifth principle.*

Here enter in certain factors and analogies that it would be of profit to us to mention at this juncture.

This fifth principle embodies the basic vibration of the fifth plane, either cosmically or systemically considered. A certain sound of the logoic Word, when it reaches the mental plane, causes a vibration in the matter of that plane, arrests its tendency to dissipate, causes it to take spheroidal form, and builds it literally into a body which is held in coherent shape by a mighty deva Entity, the Raja Lord of the mental plane. Exactly the same procedure ensued on cosmic levels, when a still mightier sound was uttered by the ONE ABOUT WHOM NAUGHT MAY BE SAID, and the utterance of this caused a vibration on the fifth cosmic plane. Certain great Entities became active, including such relatively unimportant Beings as our solar Logos and His group.

This fifth principle is the distinctive coloring of a particular group of solar Logoi on the causal level of the cosmic mental, and is the animating factor of Their Existence, the reason of Their manifesting through various solar systems, and the great Will-to-be that brings Them forth into objectivity.

Manas has been defined as mind, or that faculty of logical deduction and reasoning, and of rational activity that distinguishes man from the animals. Yet it is something much more than that for it underlies all manifestation, and the very shape of an amoeba, and the discriminative faculty of the lowest atom or cell, is actuated by mind of some kind or another. It is only as the place of that discriminating cell or atom within its greater sphere is apprehended, and recognised, that any clear conception will be gained of what that coherent rational inclusive mentality may be.

A Treatise on Cosmic Fire, pp. 309/10

• 2. *Manas is electricity.*

. . . Certain electrical phenomena distinguish a human being, only (as they have not been expressed or considered in terms of electricity) the analogy has been lost sight of. These demonstrations might be considered as:

First, that coherent VITALITY which holds the entire body revolving around the central unit of force. It must here be remembered that the entire manifestation of a solar system consists of the etheric body, and the dense body of a Logos.

Second, that radiatory MAGNETISM which distinguishes man, and makes him active in two ways:

> In relation to the matter of which his vehicles are composed.
> In relation to the units which form his group.

Third, that ACTIVITY on the physical plane which results in due performance of the will and desire of the indwelling entity, and which in man is the correspondence of the Brahma aspect.

These three electrical manifestations—vitality, magnetism, and fohatic impulse—are to be seen at work in a solar Logos, a Heavenly Man and a human being. They are the objective manifestations of the psychic nature, which (in a solar Logos, for instance) we speak of in terms of quality, and call will, wisdom, activity.

A Treatise on Cosmic Fire, p. 313

- 3. *Manas is that which Produces Cohesion.*

We come now to our third definition: The manasic principle is above all else that cohesive something which enables an Entity (whether Logos, Heavenly Man, or man) to work:

a. Through form, and thus exist.

b. By means of progressive development or cyclic evolution.

c. On certain planes, that are, for the entity concerned, the battleground of life, and the field of experience.

d. By the method of manifestation, which is a gradual growth from a dim and distant dawn through an ever increasing splendour of light to a blaze of effulgent glory; then through a steadily dimming twilight to final obscuration. Dawn, day, midday, twilight, night—thus is the order for the Logos, for a planetary Logos, and for man.

If the above four points are carefully studied, it will be found that they are fairly comprehensive, and embody the four points that are as yet the only ones available for man in this fourth round.

A Treatise on Cosmic Fire, pp. 332/3

• 4. *Manas is the Key to the Fifth Kingdom in Nature*

We might also define manas as the key to the door through which entrance is made into the fifth kingdom of nature, the spiritual kingdom. Each of the five kingdoms is entered by some one key, and in connection with the first two kingdoms—the mineral and vegetable—the key or method whereby the life escapes into the higher kingdom is so inexplicable to man as his present stage of intelligent apprehension that we will not pause to consider it. In relation to the animal kingdom it might be said that the key whereby entrance is effected into the human kingdom is that of *instinct*. . . .

Man passes into the fifth kingdom through the transmutation of the discriminative faculty of mind, which—as in the animal's individualisation—brings about at a certain stage a spiritual individualisation which is the correspondence on higher levels to what transpired in Lemurian days. Therefore, we have:

Instinct...The key from the animal into the human kingdom or from the third into the fourth kingdom.

Manas......The key from the human into the spiritual kingdom, or from the fourth kingdom into the fifth kingdom.

A Treatise on Cosmic Fire, pp. 334/5

• 5. *Manas is the Synthesis of the Five Rays*

One other definition might be given even though its abstruseness may prove but a bewilderment to the student.

Manas is the united faculty of four of the Heavenly Men, synthesised through a fifth Heavenly Man on the third plane of the system. These five Heavenly Men were the logoic embodiment in an earlier system and achieved the fullness of manasic life. Their synthetic life is that which is primarily understood when we speak of Brahma, that cosmic Entity Who is the sum-total of logoic active intelligence. For lack of better terms we call Them the Lords of the four minor Rays, Who find Their synthesis through the third Ray of Activity. They have been called in an endeavour to express the principles which They embody:

CREATION – THE BUILDING OF FORM

1. The Lord of Ceremonial Magic.
2. The Lord of Abstract Idealism, or Devotion.
3. The Lord of Concrete Science.
4. The Lord of Harmony and Art.

These four function through the fourth cosmic ether, and have vehicles of buddhic matter. They merge into the greater life of the Lord of the third Ray of Aspect on atmic levels, and these four (with the one synthetic Ray), are the totality of manasic energy. They are the life of the five lower planes. *A Treatise on Cosmic Fire*, p. 336

6. *Manas is Intelligent Will or Purpose of an Existence.*

Manas might finally be defined as the intelligent will and ordered purpose of every self-conscious entity. I would urge the student to bear in mind certain basic facts which will serve to keep his mind clear, and which will enable him to comprehend something of the place which this fire of mind holds in the cosmos and the solar system, and (needless to say), in his life also,—the reflection of the other two.

He should ever remember that manas is a *principle of the Logos*, and necessarily therefore is felt in all those evolutions which are a part of His nature but is allied especially to the throat and head centres; it is the active intelligent factor which enables a solar Logos, a planetary Logos or Heavenly Man, and a human being to:

a. Use intelligently a form or vehicle.
b. Build faculty into the causal body.
c. Reap the benefit of experience.
d. Expand the consciousness.
e. Make progress towards a specified goal.
f. Discriminate between the two poles.
g. Choose the direction in which his activity shall trend.
h. Perfect the form as well as use it.
i. Obtain control of active substance, and turn its forces into desired channels.

j. Co-ordinate the different grades of matter, and synthesise the utilised forms till each and all show a unanimous line of action and express simultaneously the will of the Indweller.

All these ends are the result of the manasic development and perhaps the student might apprehend the underlying idea more clearly if it is realised that

a. The Spirit employs *manas* in all that concerns matter, the electrical substance, or the active akasha.
b. The Spirit employs *buddhi* in all that relates to the psyche, that relates to the soul of the world, to the soul of an individual, or to the soul of every form.
c. The Spirit employs will or *atma* in all that relates to the essence of all, to itself, considering the essence and the Self as pure Spirit as distinguished from spirit-matter.

A Treatise on Cosmic Fire, pp. 337/8

7. THE SOUL

• The soul is the conscious factor in all forms, the source of that awareness which all forms register and of that responsiveness to surrounding group conditions which the forms in every kingdom of nature demonstrate.

Therefore the soul might be defined as that significant aspect in every form (made through this union of spirit and matter) which feels, registers awareness, attracts and repels, responds or denies response and keeps all forms in a constant condition of vibratory activity.

A Treatise on White Magic, pp. 36

• The soul therefore, viewed from one angle, is an aspect of the body, for there is a soul in every atom comprising all bodies in all kingdoms in nature. The subtle coherent soul which is the result of the bringing together of spirit and matter exists as an entity apart from the body nature, and constitutes (when separated from the body) the etheric body, the double, as it is sometimes called, or the counterpart of the physical body. This is the sum total of the soul of the atoms constituting the physical body. It is the true form; it is the principle of coherence in every form. *Esoteric Psychology* I, p. 54

• It should be borne in mind that the soul of matter, the anima mundi, is the sentient factor in substance itself. It is the responsiveness of matter throughout the universe and that innate faculty in all forms, from the atom of the physicist, to the solar system of the astronomer, which produces the undeniable intelligent activity which all demonstrate. It can be called attractive energy, coherency, sentiency, aliveness, awareness or consciousness, but perhaps the most illuminating term is that the soul is the quality which every form manifests. It is that subtle something which distinguishes one element from another, one mineral from another. It is the intangible essential nature of the form which in the vegetable kingdom determines whether a rose or a cauliflower, an elm or a watercress shall come into being; it is a type of energy which distinguishes the varying species of the animal kingdom and makes one man different from another in his appearance, nature and character. The scientist has tabulated, investigated and analysed the forms; names have been selected and given to the elements, and the minerals, the forms of vegetable life and the varying species of animals; the structure of the forms and the history of their evolutionary progress have been studied and deductions and conclusions have been reached, but the solution of the problem of life itself still eludes the wisest, and until the understanding of the "web of life" or of the body of vitality which underlies every form and links every part of a form with every other part is recognised and known to be a fact in nature, the problem will remain unsolved. . . .

1. The soul, macrocosmic and microcosmic, universal and human, is that entity which is brought into being when the spirit aspect and the matter aspect are related to each other. . . .

2. The soul is the attractive force of the created universe and (when functioning) holds all forms together so that the life of God may manifest or express itself through them. . . .

3. This soul manifests differently in the various kingdoms of nature, but its function is ever the same, whether we are dealing with an atom of substance and its power to preserve its identity and form, and carry forward its activity along its own lines, or whether we deal with a form in one of the three kingdoms of nature, held coherently together, demonstrating characteristics, pursuing its own instinctual life and working as a whole towards something higher and better. . . .

4. The qualities, vibrations, colours, and characteristics in all the kingdoms of nature are soul qualities, as are the latent powers in any form seeking expression, and demonstrating potentiality. In their sum total at the close of the evolutionary period, they will reveal what is the nature of the divine life and of the world soul,—that oversoul which is revealing the character of God. . . .

5. The soul of the universe is—for the sake of clarity—capable of differentiation or rather (owing to the limitations of the form through which that soul has to function) capable of recognition at differing rates of vibration and stages of development. The soul nature in the universe therefore manifests in certain great states of awareness with many intermediate conditions, of which the major can be enumerated as follows:

a. Consciousness, or that state of awareness in matter itself, . . .

b. Intelligent sentient consciousness, i.e. that evidenced in the mineral and vegetable kingdoms. . . .

c. Animal consciousness, the awareness of soul response of all forms in the animal kingdom, producing their distinctions, species and nature.

d. Human consciousness, or self-consciousness, towards which the development of the life, form and awareness in the other three kingdoms has gradually tended. . . .

e. Group consciousness, which is the consciousness of the great sum totals, . . . There are stages in this realisation, mounting all the way from that tiny group recognition of the probationary disciple up to the completed group awareness of the life in Whom all forms have their being, the consciousness of the planetary Logos, . . .

The soul therefore may be regarded as the unified sentiency and the relative awareness of that which lies back of the form of a planet and of a solar system.

A Treatise on White Magic, pp. 33/6, 37/9

• What is the meaning of the following words: Sentiency; Consciousness or Awareness; The Energy of Light?

We shall now consider our last question, and I shall indicate to you, in general terms,—necessarily limited by the inadequacy of language,—the significance of the outstanding soul qualities:

a. Sentiency, or sensitive response to contact, and by that means the subsequent growth in knowledge.

b. Consciousness, awareness of environment, and the development of instruments whereby consciousness may be increasingly developed.

c. Light, or radiation, the effect of the interplay between the life and the environment.

The first point that I seek to make is a difficult one to grasp for those beneath the rank of initiate or accepted disciple of the higher stages. The soul is that factor in matter (or rather that which emerges out of the contact between spirit and matter) which produces sentient response and what we call consciousness in its varying forms; it is also that latent or subjective essential quality which makes itself felt as light or luminous radiation. It is the "self-shining from within" which is characteristic of all forms. Matter, per se, and in its undifferentiated state, prior to being swept into activity through the creative process, is not possessed of soul, and does not therefore possess the qualities of response and of radiation. Only when,—in the creative and evolutionary process,—these two are brought into conjunction and fusion does the soul appear and give to these two aspects of divinity the opportunity to manifest as a trinity and the chance to demonstrate sentient activity and magnetic radiatory light. As all that we shall posit in this treatise is to be approached from the angle of human evolution, it might be stated that only when the soul aspect is dominant is the response apparatus (the form nature of man) fulfilling its complete destiny, and only then does true magnetic radiation and the pure shining forth of light become possible. Symbolically, in the early stages of human evolution, man is, from the angle of consciousness, relatively unresponsive and unconscious, as is matter in its early stages in the formative process. The achievement of full awareness is of course the goal of the evolutionary process. Again symbolically speaking, the unevolved man emits or manifests no light. The light in the head is invisible, though the clairvoyant investigators would see the dim glow of the light within the elements which constitute the body, and the light hidden in the atoms which constitute the form nature.

As evolution proceeds, these dim points of "dark light" intensify their glow; the light within the head flickers at intervals during the life of the average man, and becomes a shining light as he enters upon the path of discipleship. When he becomes an initiate, the light of the atoms is so bright, and the light in the head so intense (with a paralleling stimulation of the centres of force in the body), that the light body appears. Eventually this body of light becomes externalised and of greater prominence than the dense tangible physical body. This is the body of light in which the true son of God consciously dwells. After the third initiation, the dual light becomes accentuated and takes on a still greater brilliancy through the blending with it of the energy of spirit. This is not really the admission or the re-combining of a third light, but the fanning of the light of matter and the light of the soul into a greater glory through the Breath of the spirit. Something anent this light has been earlier indicated in A Treatise on Cosmic Fire. Study it and seek to understand the significance of this process. In the understanding of these aspects of light comes a truer perspective as to the nature of the fires in the human expression of divinity.

It must never be forgotten that the soul of all things, the anima mundi, as it expresses itself through all the four kingdoms in nature, is that which gives to our planet its light in the heavens. The planetary light is the sum total of the light, dim and uncertain, to be found in all atoms of radiatory and vibratory matter or substance, which compose all forms in all kingdoms. Added to this, there is, within the planet and also within each kingdom in nature, the correspondence to the

CREATION – THE BUILDING OF FORM

etheric body with its centres of radiant energy, found underlying or "substanding" the outer physical form. Man's etheric body is a corporate part of the planetary etheric body, and constitutes its most refined and most highly developed aspect. As the aeons pass away there is a growing intensity of light radiating forth from our planet. This does not necessarily mean that a dweller on Neptune would see our planet glowing with an increasingly brighter light, though this does happen in a few cases in the universe. But it means that, from the standpoint of a clairvoyant vision, the etheric planetary body will grow in vivid radiation and glory as that radiation expresses more and more the true light of the soul.

The soul is light essentially, both literally from the vibratory angle, and philosophically from the angle of constituting the true medium of knowledge. The soul is light symbolically, for it is like the rays of the sun, which pour out into the darkness; the soul, through the medium of the brain, causes revelation. It throws its light into the brain, and thus the way of the human being becomes increasingly illumined. The brain is like the eye of the soul, looking out into the physical world; in the same sense the soul is the eye of the Monad, and in a curious and occult sense, the fourth kingdom in nature constitutes on our planet the eye of the planetary Deity.

Esoteric Psychology I, pp. 129/32

IX. THE GREATER AND LESSER BUILDERS

• God, Ray, Life, and Man are all psychological entities and builders of forms. Therefore a great psychological life is appearing through the medium of a solar system.

Esoteric Psychology I, p. 21

• I have divided the groups of devas and elementals into evolutionary and involutionary Builders—those who are in themselves positive force, and those which are negative force, the conscious and the blind workers. It is absolutely essential that students bear in mind here that we are studying the mystery of electricity and therefore must remember the following facts:—

a. *Introductory Remarks.*

The Mystery of Electricity. The greater Builders are the positive aspect of substance or of electrical phenomena whilst the lesser Builders are the negative aspect.

Two types of force are represented in the activities of these two groups and it is their interaction and interplay which produces Light, or the manifested solar system.

Their sumtotal is substance in its totality, the intelligent active form, built for the purpose of providing a habitation for a central subjective life.

They are also the sumtotal of the Pitris, or Fathers of mankind, viewing mankind as the race itself, the fourth kingdom in nature, the Heavenly Men *in physical manifestation.* This is a most important point to emphasise. These deva activities in relation to Self-Consciousness (which is the distinctive characteristic of humanity) can best be studied in the large, or through the consideration of groups, of races, and of the life of the scheme, the manifestation of one of the Heavenly Men. When the student brings his study of deva work down to the terms of his own individual life he is apt to become confused through too close a juxtaposition.

THE GREATER AND LESSER BUILDERS 103

The greater Builders are the solar Pitris, whilst the lesser Builders are the lunar ancestors. I would here explain the occult meaning of the word "ancestor," as used in esotericism. It means literally initiatory life impulse. It is that subjective activity which produces objectivity, and concerns those emanatory impulses which come from any positive centre of force, and which sweep the negative aspect into the line of that force, and thus produce a form of some kind. The word "ancestor" is used in connection with both aspects.

The solar Logos is the initiatory impulse or Father of the Son in His physical incarnation, a solar system. He is the sumtotal of the Pitris, in the process of providing *form*. The union of Father (positive force) and Mother (negative force) produces that central blaze which we call the form, the body of manifestation of the Son. *A Heavenly Man* holds an analogous position in relation to a planetary scheme. He is the central germ of positive life or force, which, in due course of time, demonstrates as a planetary scheme, or an incarnation of the planetary Logos. A *man* similarly is the positive life or energy which, through action on negative force, creates bodies of manifestation through which he can shine or radiate.

The lesser Builders are the negative aspect and are swept into action in group formation through the play of positive force upon them, or through the action of the conscious Minds of the system. At the present stage of evolution—during the period of Light—it is difficult for the human being (until he has attained the consciousness of the Ego) to differentiate between the types of force, and to work *consciously* with these dual aspects. An Adept of the Light works with force in substance, viewing substance as that which is negative, and therefore occultly to be moved, and He can do this because He has (in the three worlds of His endeavour) achieved unity, or the point of balance and equilibrium, and can therefore balance forces and deal with positive and negative energies as appears best in the interests of the plan of evolution. The Brother of Darkness, knowing himself to be positive force in essence, works with negative substance, or with the lesser Builders to bring about ends of his own, incited thereto by selfish motive. The Brothers of Light co-operate with the positive aspect in, and of, all forms—the building devas of evolutionary intent—in order to bring about the purposes of the Heavenly Man Who is the sumtotal of planetary physical manifestation.

It can be seen, therefore, how necessary it is that the functions of the devas of all grades be comprehended. It is however equally important that man should refrain from the manipulation of these forces of nature until such time as he "knows" himself, and his own powers, and until he has fully unfolded the consciousness of the ego; then, and only then, can he safely, wisely, and intelligently co-operate in the plan. As yet, for the average man or even the advanced man this is dangerous to attempt and impossible to accomplish. . . .

The building devas are the Ah-hi, or Universal Mind. They contain within their consciousness the plan logoic, and inherently possess the power to work it out in time and space, being the conscious forces of evolution.

They not only embody the Divine Thought but are that through which it manifests, and its actuating activity. They are essentially motion. The lesser builders are more particularly the material form which is actuated, and in their cohorts are the substance of matter (considering substance as that which lies back of matter).

They are that which produces concretion and which gives form to the abstract. The terms "rupa" and "arupa" devas are relative, for the formless levels and the formless lives are only so from the standpoint of man in the three worlds; the formless lives are those which are functioning in and through the etheric body of the Logos, formed of the matter of the four higher planes of the system. . . .

The lunar Pitris, the builders of man's lunar body and their correspondence in the other kingdoms of nature, are the sumtotal of the dense physical body of the Logos, or the substance of the mental, astral and physical planes (the gaseous, liquid and dense bodies which form a unity, His physical vehicle, viewing it apart from the etheric). They are the product of an earlier solar system; their activities date from there. That system stands to the present one as the lunar chain to ours. . . .

THE GREATER AND LESSER BUILDERS 105

In the first solar system the negative substance aspect, the Mother aspect or matter, was perfected. The lower Pitris dominated. In this system force activity lies in the hands of the solar Pitris or greater devas. At the close of the mahamanvantara they will have built according to the plan a perfect sheath or vehicle of expression for the Divine Thought, and this through the manipulation of negative substance; they utilise the heat of the Mother to nourish the germ of the Divine Thought, and to bring it to fruition. When the germ has developed to maturity the Mother aspect no longer has a place, and the Man occultly is freed or liberated. This idea runs through all manifestations, and the kingdoms of nature or the form (no matter what form it may be) nourish the germ of that which is the next step on in the evolutionary process, and are considered the Mother aspect. This aspect is eventually discarded and superseded. For example, the third kingdom, the animal, in the early stages nourishes and preserves the germ of that which will some day be a man; the personality is the preserver of that which will some day unfold into spiritual man.

It will thus become apparent to students how the Heavenly Man, viewing Him as a solar Deity, a self-conscious Entity, works with His negative aspect through positive force, from logoic etheric levels upon the three aspects of the logoic dense physical, thus bringing to maturity the atoms and cells of His Body, fostering the germ of self-consciousness, fanning the flame until each unit becomes fully group conscious and aware of his place within the body corporate. Each human being likewise, functioning in the three worlds, works in a corresponding way upon the conscious cells of his bodies, until each atom eventually achieves its goal. The Heavenly Man works necessarily through egoic groups, pouring positive force upon them until they emerge from passivity and negativity into potency and activity. Man works correspondingly through his centres upon his sheaths, and has a responsibility to the lesser lives which under the karmic law must be worked out. This is the basis of the evolutionary process.

A Treatise on Cosmic Fire, pp. 612/6, 618/9

106 *THE GREATER AND LESSER BUILDERS*

1. THE ENTIFYING OF THE SOLAR SPHERE

ETHERIO – ATOMIC PHILOSOPHY OF FORCE

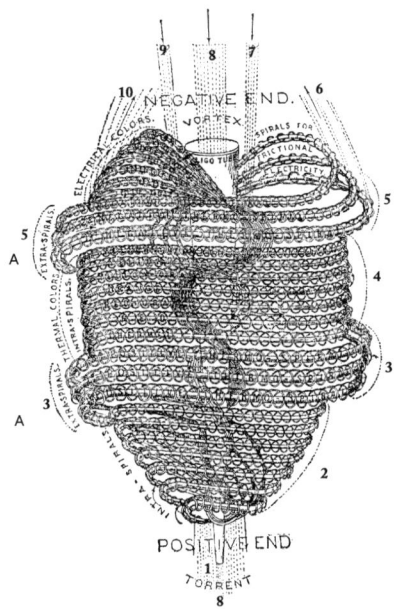

The general Form of an Atom, including the spirals and 1st Spirillae, together with influx and efflux ethers, represented by dots, which pass through these spirillae. The 2nd and 3rd spirillae with their still finer ethers are not shown.

(EDWIN D. BABBITT, *Principles of Light and Color,* NEW YORK, 1878.)

• This solar sphere is closely similar to the atom pictured in the book by Babbitt and later in *Occult Chemistry* by Mrs. Besant. The Sun is heart-shaped, and (seen from cosmic angles) has a depression at what we might call its north pole. This is formed by the impact of logoic energy upon solar substance.

This energy which impinges upon the solar sphere, and is thence

THE GREATER AND LESSER BUILDERS

distributed to all parts of the entire system, emanates from three cosmic centres and, therefore, is triple during this particular cycle.

a. From the sevenfold great Bear.
b. From the Sun Sirius.
c. From the Pleiades.

It must be remembered that the possible cosmic streams of energy available for use in our solar system are seven in number, of which three are major. These three vary during vast and incalculable cycles.

Students may find it of use to remember that,
a. The Law of Economy demonstrates as an urge,
b. The Law of Attraction as a pull,
c. The Law of Synthesis as a tendency to concentrate at a centre, or to merge.

The streams of energy which pour forth through the medium of the Sun from the egoic lotus and which are in reality "logoic Soul energy" attract to them that which is akin to them in vibration. This may sound rather like the statement of a platitude, but is susceptible of really deep significance to the student, being accountable for all systemic phenomena. These streams pass in different directions, and in the knowledge of occult direction comes knowledge of the various hierarchies of being, and the secret of the esoteric symbols.

The main stream of energy enters at the top depression in the solar sphere and passes through the entire ring-pass-not, bisecting it into two halves.

With this stream enters that group of active lives whom we call the "Lords of Karma." They preside over the attractive forces, and distribute them justly. They enter, pass to the centre of the sphere and there (if I may so express it) locate, and set up the "Holy Temple of Divine Justice," sending out to the four quarters of the circle the four Maharajahs, their representatives. So is the equal armed Cross formed—and all the wheels of energy set in motion. This is conditioned by the karmic seeds of an earlier system, and only that substance is utilised by the Logos, and only those lives come into manifestation who have set up a mutual attraction.

These five streams of living energy (the one and the four) are the basis of the onward march of all things; these are sometimes esoterically called "the forward moving Lives." They embody the *Will* of the Logos. It is the note they sound and the attractive pull which they initiate which bring into contact with the solar sphere a group of existences whose mode of activity is spiral and not forward.

These groups are seven in number and pass into manifestation through what is for them a great door of Initiation. In some of the occult books, these seven groups are spoken of as the "seven cosmic Initiates Who have passed within the Heart, and there remain until the test is passed." These are the seven Hierarchies of Beings, the seven Dhyan Chohans. They spiral into manifestation, cutting across the fourfold cross, and touching the cruciform stream of energy in certain places. The places where the streams of love energy cross the streams of will and karmic energy are mystically called the "Caves of dual light" and when a reincarnating or liberated jiva enters one of these Caves in the course of his pilgrimage, he takes an initiation, and passes on to a higher turn of the spiral.

Another stream of energy follows a different route, which is a little difficult to make clear. This particular set of active lives enter the heart shaped depression, pass around the *edge* of the ring-pass-not to the lowest part of the solar sphere and then mount upwards, coming into opposition therefore with the stream of downpouring energy. This stream of force is called "lunar" force for lack of a better term. They form the body of the raja Lord of each of the planes, and are governed by the Law of Economy.

All these streams of energy form geometrical designs of great beauty to the eye of the initiated seer. We have the transverse and bisecting lines, the seven lines of force which form the planes, and the seven spiralling lines, thus forming lines of systemic latitude and longitude, and their interplay and interaction produce a whole of wondrous beauty and design. When these are visualised in colour, and seen in their true radiance, it will be realised that the point of attainment of our solar Logos is very high, for the beauty of the logoic Soul is expressed by that which is seen.

A Treatise on Cosmic Fire, pp. 1181/4

THE GREATER AND LESSER BUILDERS 109

a. THE HIERARCHIES

• If the student bears in mind that the nature of the form is dependent upon the *quality* of the incarnating Life, he will have also to bear in mind the distinctions between the various groups of Hierarchies, for the Lives in those groups are of a quality diverse to each other and the forms through which they manifest are equally distinct and diverse. Therefore, we must distinguish between:

1. The involutionary groups.
2. The evolutionary groups.
3. The seven groups of lives which we call the lunar Fathers:
    a. Three incorporeal who are the elemental kingdoms.
    b. Four material who are the forms of the four kingdoms on the upward arc.
4. The seven hierarchies of Lives.
5. The seven groups of solar Angels.

A Treatise on Cosmic Fire. p. 1194

• There must not be confusion as to the distinction between the hierarchies of Beings and the seven Rays, for though there is close connection, there is no resemblance. The "Rays" are but the primordial forms of certain Lives who "carry in their Hearts" all the Seeds of Form. The Hierarchies are the manifold groups of lives, at all stages of unfoldment and growth who will use the forms. The Rays are vehicles and are, therefore, negative receivers. The Hierarchies are the users of the vehicles, and it is the nature of these lives and the quality of their vibration which under this great Law of Attraction brings to them the needed forms. These are the two primal distinctions, Life and Form, and these two are the "Son of God," the second Person of the Trinity in His form-building aspect. They are the Builders and equally exist in three groups with their lesser differentiations. It is not necessary here to place these groups on certain planes in the solar system.

These hierarchies of Beings Who come in on the Ray of Light from the centre are the seeds of all that later is and it is only as they pass out into manifestation and the forms which they are to occupy are gradually evolved, that consideration of the planes becomes necessary. The planes are to certain of these hierarchies what the sheaths of the Monad are to it; they are veils for the Life indwelling; they are media of expression, and exponents of force or energy of a specialised kind. The quality of a Ray is dependent upon the quality of the hierarchy of Beings who use it as a means of expression. These seven hierarchies are veiled by the Rays, but each is found behind the veil of every ray, for in their totality they are the informing lives of every planetary scheme within the system; they are the life of all interplanetary space, and the existences who are expressing themselves through the planetoids, and all forms of lesser independent life than a planet. Let me briefly give certain hints concerning these hierarchies which may serve to elucidate that contained in the *Secret Doctrine* concerning them.

What is here imparted is not in itself new, but is the synthesising of much already known and its gathering together in the form of brief enunciated facts.

Each of these seven hierarchies of Beings Who are the Builders or the *Attractive* Agents are (in their degree) intermediaries; all embody one of the types of force emanating from the seven constellations. Their intermediary work, therefore, is dual:

1. They are the mediators between Spirit and matter.
2. They are the transmitters of force from sources extraneous to the solar system to forms within the solar system.

Each of these groups of beings is likewise septenary in nature, and the forty-nine fires of Brahma are the lowest manifestation of their fiery nature. Each group also may be regarded as "fallen" in the cosmic sense, because involved in the building process, or the occupiers of forms of some degree of density or another.

A Treatise on Cosmic Fire, pp. 1194/6

• I would only point out that just as we teach in the occult wisdom that there is a definite progression from one kingdom into the next higher above, so there is a similar activity in the realm of the hierarchies. The lives which compose a Hierarchy pass in ordered cycles

THE GREATER AND LESSER BUILDERS

into the next above, though the word "above" but serves to mislead. It is *consciousness* and realisation which must be considered as being transferred and the consciousness of one hierarchy expands into that of the next higher.

This can also be viewed in terms of energy. The negative lives of a hierarchy follow the following sequence:

1. Negative energy.
2. Equilibrised energy.
3. Positive energy.

The positive lives of one hierarchy become the negative lives of another when they pass into it, and this it is which leads to the general confusion of ideas under which the average student labours. If he is to comprehend the matter with accuracy, he must study each hierarchy in a threefold manner, and view it also in its transitional state, as the negative blends and merges into the positive, and the positive becomes the negative pole of a higher vibratory stage. There are, therefore, nine states of consciousness through which each hierarchy has to pass, and some idea of the significance of this and their relativity can be gained by a consideration of the nine Initiations of the fourth Creative Hierarchy. Within these nine distinct expansions through which each life in each hierarchy must pass, are to be found lesser expansions and it is here that the main difficulty for the student of divine psychology lies. The whole subject concerns the psyche, or second aspect, of every life—superhuman, human and subhuman—and only when the true psychology is better understood will the subject take on its true importance. Then the nine unfoldments of each hierarchy will be somewhat comprehended, and their relative importance assigned. *A Treatise on Cosmic Fire,* pp. 1208/9

See enumeration and description of the Hierarchies:
A Treatise on Cosmic Fire, pp. 1196/1211

2. AGNI–THE RULER OF FIRE

• We proceed now to take up the consideration of the *Ruler of Fire, AGNI,* and are brought to the study of the vitality that energises and the Life that animates; to the contemplation of the Fire that drives, propels, and produces the activity and organisation of all forms. The realisation of this will reveal the fact that what we are dealing with is the "Life and the lives," as it is called in the *Secret Doctrine;* with Agni, the Lord of Fire, the Creator, the Preserver, and the Destroyer; and with the forty-nine fires through which He manifests. We are dealing with solar fire per se, with the essence of thought, with the coherent life of all forms, with the consciousness in its evolving aspect, or with Agni, the sumtotal of the Gods. He is Vishnu and the Sun in His glory; He is the fire of matter and the fire of mind blended and fused; He is the intelligence which throbs in every atom; He is the Mind that actuates the system; He is the fire of substance and the substance of the fire; He is the Flame and that which the Flame destroys.

Students of the *Secret Doctrine* when they read carelessly are apt to consider Him only as the fire of matter and omit to note that He is Himself the sumtotal—and this is especially the case when they find that Agni is the Lord of the mental plane. He is the animating life of the solar system, and that life is the life of God, the energy of the Logos, and the manifestation of the radiance which veils the Central Sun. Only as He is recognised as Fohat, the energy of matter, as Wisdom, the nature of the Ego and its motivation, and as essential unity, can any due conception be arrived at as to His nature or being. He is not the solar Logos on the cosmic mental plane, for the egoic consciousness of the Logos is more than His physical manifestation, but *Agni is the sumtotal of that portion of the logoic Ego which is reflected down into His physical vehicle; He is the life of the logoic Personality, with all that is included in that expression.* He is to the solar Logos on His own plane what the coherent personality of a human being is to his Ego in the causal body. This is a very important point to be grasped, and if meditated upon will bring to the student much enlightenment. His is the life that fuses and blends the threefold nature of the Logos when in physical incarnation; His is the coherent force that makes a unity of the triple logoic Personality, but man can only arrive at His essential nature by the study of the logoic physical vehicle—hence the difficulty; he can only understand by a consideration of His psychic

THE GREATER AND LESSER BUILDERS 113

emanation as it can be sensed and viewed by passing the history of the races in retrospect. Man's personality reveals his nature as his life progresses; his psychic quality unfolds as the years slip away, and when he passes out of incarnation he is spoken of in terms of quality, good or bad, selfish or unselfish; the effect of his "emanation" during life is that which remains in men's minds. Thus only can the logoic personality express itself, and our knowledge of His nature is consequently limited by our close perspective, and handicapped by the fact that we are participants in His life, and integral parts of His manifestation.

It is only as we begin to function upon the buddhic plane that we can in any way "live in the subjective" side of nature, and it is only as our knowledge of the spiritual life increases, and as we pass definitely through the portal of initiation into the fifth kingdom that we can appreciate the distinction between the dense physical, and the vital body. Only as we become polarised in the cosmic etheric body and are no longer held prisoner by a dense material sheath (for the three lower planes are but the dense body of the Logos) do we come to a fuller understanding of the psychic nature of the Logos, for we stand then in the body which bridges the gulf between the dense physical, and the astral body of the Logos. Only when this is the case do we understand the function of the Lord Agni as the vital life of the cosmic etheric, as the vitality of the Heavenly Men and the activity of Their sheaths.

A Treatise on Cosmic Fire, pp. 601/4

• Some thoughts on FIRE. . . . Most of the psychological phenomena of the earth are—as you will realise, if you think clearly,— under the control of the Deva Lord Agni, the great primary Lord of Fire, the Ruler of the mental plane. Cosmic fire forms the background of our evolution; the fire of the mental plane, its inner control and dominance and its purifying asset coupled to its refining effects, is the aim of the evolution of our three-fold life. When the inner fire of the mental plane and the fire latent in the lower vehicles merge with the sacred fire of the Triad the work is completed, and the man stands adept. The at-one-ment has been made and the work of aeons is completed. All this is brought about through the co-operation of the Lord

Agni, and the high devas of the mental plane working with the Ruler of that plane, and with the Raja-Lord of the second plane.

Letters on Occult Meditation, p. 100

3. THE GREATER BUILDERS– FIRE DEVAS OF THE PHYSICAL PLANE

• The subject of our consideration now is the fire devas of the physical plane, those great building devas who are working out the purposes of the Logos in his dense, physical body. Let us get our ideas as clear as possible on this matter; in the following tabulation, the status of these devas will be apparent at a glance:

Name	Cosmic Plane	Systemic Plane	Nature	Ruler
Agnichaitan	7th subplane cosmic physical	Physical	Densest concretion	Kshiti
Agnisuryan	6th subplane cosmic physical	Astral	Liquid	Varuna
Agnishvatta	5th subplane cosmic physical	Mental	Gaseous	Agni

The Agnichaitans. These are the devas who construct, and build in matter of the densest kind in connection with logoic manifestation. They function on the seventh subplane of the cosmic physical plane, and are the producers of the greatest concretion. In the planetary body of our planetary Logos they are the builders of the Earth, His densest form, and throughout the entire solar system they are the sum-total of that activity and vibration which demonstrates through what we call "solid substance."

Therefore, it will be apparent that under the law they will have a peculiarly powerful effect on the lowest subplane of the systemic physical plane; hence their esoteric appellation of the "Agnichaitans of the inner or central heat." They are the totality of the lowest vibration in the cosmic physical vehicle.

THE GREATER AND LESSER BUILDERS 115

The Agnisuryans are the builders on the sixth subplane of the cosmic physical plane, our systemic astral plane. They represent, as I have before hinted, the sympathetic nervous system in the logoic physical body, just as their brothers of the seventh vibration represent the sumtotal of the circulatory or blood system. A hint to the student who is interested in the physiological key lies in the relationship between the two great groups of devas who build and construct the most objective portion of logoic manifestation, and the two groups of corpuscles which in their mutual interaction hold the body in health; there is an analogy also in the relationship between the devas of the astral plane, and the motor and sensory nerves of the physical body. I will not enlarge upon this angle of vision.

These devas have to do, in a very esoteric sense, with the nerve plexus in the:

a. Solar system. (Physical Sun)
b. Planetary scheme. (Dense Planet)
c. Human physical body. (Dense Body)

and are therefore a powerful factor in the eventual vitalisation of the centres in man. The etheric centres, or the focal points of force of a Heavenly Man are on the fourth cosmic ether, the buddhic plane. The astral plane is closely allied to the buddhic, and as the etheric centres of our Heavenly Man, for instance, come into full activity, the force is transmitted through the astral correspondence to the fourth physical ether, in which the centres of man exist.

The Agnishvattas are the builders on the fifth or gaseous subplane of the cosmic physical, and—from the human standpoint—are the most profoundly important, for they are the builders of the body of consciousness per se. From the psychic standpoint of occult physiology, they have a close connection with the physical brain, the seat or empire of the Thinker, and as at this stage all that we can know must be viewed kama-manasically, it will be apparent that between the sympathetic nervous system and the brain is such a close interaction as to make one organised whole. This microcosmic correspondence is of interest, but in studying these groups of devas at present we will view them principally in their work as systemic and planetary builders, leaving the student to trace out for himself the human analogy.

A Treatise on Cosmic Fire, pp. 633/5

4. THE TRANSMITTING DEVAS

• It will have been noted that in the enumeration of these two main groups (Greater and Lesser devas), we did not touch upon that great group of Builders who are called esoterically "Those who transmit the Word.". . . The "Transmitters of the Word" upon the first subplane or atomic level are those who take up the vibratory sound as it reaches them from the astral plane and—passing it through their bodies—send it forth to the remaining subplanes. These transmitters may be, for purposes of clarity, considered as seven in number. In their totality they form the atomic physical bodies of the Raja Lord of the plane and in a peculiarly occult sense these seven form (in their lower differentiations on etheric levels) the sumtotal of the etheric centres of all human beings, just as on the cosmic etheric levels are found the centres of a Heavenly Man.

The connection between the centres and etheric substance, systemic and human, opens up a vast range for thought. The "Transmitters of the Word" on the atomic subplane of each plane are devas of vast power and prerogative who may be stated to be connected with the Father aspect, and embodiments of electric fire. They are all fully self-conscious, having passed through the human stage in earlier kalpas. They are also corporate parts of the seven primary head centres in the body of a solar Logos or planetary Logos.

Though connected with the Father aspect, they are nevertheless part of the body of the Son, and each of them, according to the plane which he energises, is a component part of one or other of the seven centres, either solar or planetary—planetary when only the particular centre is concerned, systemic when that centre is viewed as an integral part of the whole.

Each of these great lives (embodying deva energy of the first degree) is an emanation from the central spiritual sun in the first instance and from one of the three major constellations in the second instance. Systemically they fall into three groups: Group 1 includes those transmitters of the Word who are found on the three lower subplanes of the plane Adi, or the logoic plane. Group 2 comprises those great builders who transmit the Word on the three next systemic planes, the monadic, the atmic and the buddhic. Group 3 is formed of those who carry on a similar function in the three worlds of human endeavour. Fundamentally they are also emanations from one of the seven stars of the Great Bear in the third instance.

A Treatise on Cosmic Fire, pp. 919/20

5. THE LESSER BUILDERS–THE FIRE ELEMENTALS

• *Introductory*

We have dealt briefly with the Builders on the evolutionary arc, the greater entities who either have passed through the human kingdom, and therefore have left that stage of evolution behind them in earlier cycles, or are at this time the "solar agents" of human manifestation. All these forms of divine existence represent—in their own place—aspects of *positive force*. We come now to the consideration of the lesser builders in the three worlds, those who represent the negative aspect of force, being on the involutionary arc, and who are, therefore, the recipients of energy and influences. They are worked upon by energy, and through the activity of the greater Builders are forced into different directions in space, being built into the differing forms. The energy that works upon them, as is well known, emanates from the second aspect, and in their totality they form the great Mother.

I would call to the attention of all students the fact that these lesser builders are literally a "sea of fire" upon which the great breath, or the AUM, takes effect. Each fiery spark, or atom, becomes (through the action of the Word), vitalised with new life, and impregnated with a different type of energy. In the union of the life of atomic substance itself with that which causes the atoms to cohere, and to form vehicles of some kind or another, can be seen demonstrating the "Son of God."

Herein lies the essential duality of all manifestation; this duality is later supplemented by the life of the One Who sounds the Word. Thus is the cosmic incarnation brought about with the three factors entering in. *A Treatise on Cosmic Fire*, pp. 887/8

a. *Physical Plane Elementals*

• The elemental groups of the dense physical plane who are swept into activity by the builders, are three in number:

a. The gaseous elementals.
b. The liquid elementals.
c. The strictly dense elementals.

One group concerns itself with the fiery channels, with the fires of the human body, and with the different gases to be found within the human periphery. Another group is to be seen working in connection with the circulatory system, and with all the liquids, juices, and waters of the body; whilst the third is largely involved in the construction of the frame, through the right apportioning of the minerals and chemicals. *A Treatise on Cosmic Fire*, pp. 944/5

1) Elementals of Densest Matter

• These are the workers and builders which are concerned with the tangible and objective part of all manifestation. In their totality they literally form that which can be touched, seen, and contacted physically by man. In considering these matters we must never dissociate the various groups in our minds in a too literal sense, for they all interpenetrate and blend, in the same manner as man's physical body is compounded of dense, liquid, gaseous, and etheric matter. Diversity, producing a unity, is everywhere to be seen; this fact must constantly be borne in mind by the occult student when studying the subhuman forms of existence. There is a distinct danger in all tabulations, for they tend to the forming of hard and fast divisions, whereas unity pervades all.

Among the manipulating devas of the lowest level of the dense

THE GREATER AND LESSER BUILDERS 119

physical plane are to be found certain subterranean forms of existence, of which hints are to be found in the ancient and occult books. There is to be found in the very bowels of the earth, an evolution of a peculiar nature, with a close resemblance to the human. They have bodies of a peculiarly gross kind, which might be regarded as distinctly physical as we understand the term. They dwell in settlements, or groups, under a form of government suited to their needs in the central caves several miles below the crust of the earth. Their work is closely connected with the mineral kingdom, and the "agnichaitans" of the central fires are under their control. Their bodies are constituted so as to stand much pressure, and they are not dependent upon as free a circulation of air as man is, nor do they resent the great heat to be found in the earth's interior. Little can here be communicated anent these existences, for they are connected with the lesser vital portions of the physical body of the planetary Logos, finding their microcosmic correspondence in the feet and legs of a man. They are one of the factors which make possible the revolutionary progressive activity of a planet.

Allied with them are several other groups of low class entities, whose place in the scheme of things can only be described as having relation to the grosser planetary functions. Little is gained by enlarging upon these lives and their work; it is not possible for man in any way to contact them, nor would it be desirable. When they have pursued their evolutionary cycle, they will take their place in a later cycle in the ranks of certain deva bodies that are related to the animal kingdom.

It is commonly supposed that all the fairies, gnomes, elves, and like nature spirits are to be found solely in etheric matter, but this is not so. They are to be found in bodies of gaseous and liquid substance likewise, but the mistake has arisen for the reason that the basis of all that which can be objectively seen is the etheric structure, and these little busy lives frequently protect their dense physical activities through the agency of glamour, and cast a veil over their objective manifestation. When etheric vision is present then they can be seen, for the glamour, as we understand it, is only a veil over that which is tangible.

Students must at this juncture remember that all dense physical forms, whether of a tree, an animal, a mineral, a drop of water, or a precious stone, are in themselves elemental lives constructed of living substance by the aid of living manipulators, acting under the direction of intelligent architects. It will immediately become apparent why it is not possible in any way to tabulate in connection with this particular lowest group. A beautiful diamond, a stately tree, or a fish in the water are but devas after all. It is the recognition of this essential livingness which constitutes the basic fact in all occult investigation, and is the secret of all beneficent magic. It is not my purpose, therefore, to deal more specifically with these lowest forms of divine life, except to impart two facts, and thus give indication of the solution of two problems which have oft disturbed the average student; these are, first, the problem as to the purpose of all reptilian life, and, secondly, the specific connection of the bird evolution with the deva kingdom.

The secret of the *reptile kingdom* is one of the mysteries of the second round, and there is a profound significance connected with the expression "the serpents of wisdom" which is applied to all adepts of the good law. The reptile kingdom has an interesting place in all mythologies, and all ancient forms of truth impartation, and this for no arbitrary reason. It is not possible to enlarge upon the underlying truth which is hidden in the karmic history of our planetary Logos, and is revealed as part of the teaching given to initiates of the second degree.

The second great life impulse, or life wave, initiated by our planetary Logos, when brought in conjunction with the first, was the basis of that activity which we call evolutionary energy; it resulted in a gradual unrolling, or revelation, of the divine form. The heavenly serpent manifested, being produced out of the egg, and began its convolutions, gaining in strength and majesty, and producing through its immense fecundity millions of lesser "serpents." The reptile kingdom is the most important part of the animal kingdom in certain aspects, if such an apparently contradictory statement can be made. For all animal life can be seen passing through it during the prenatal stage, or returning to it when the form is in advanced decomposition. The connection is not purely a physical one, but it is also psychic. When the real nature and method of the kundalini, or serpent fire, is known, this relation will be better understood, and the history of the second round assume a new importance.

The secret of life lies hidden in the serpent stage,—not the life of the Spirit, but the life of the soul, and this will be revealed as the "serpent of the astral light" is truly approached, and duly studied. One of the four Lipika Lords, Who stand nearest to our planetary Logos, is called "The Living Serpent," and His emblem is a serpent of blue with one eye, in the form of a ruby, in its head. Students who care to carry the symbology a little further can connect this idea with the "eye of Shiva" which sees and knows all, and records all, as does the human eye in lesser degree; all is photographed upon the astral light, as the human eye receives impressions upon the retina. The same thought is frequently conveyed in the Christian Bible, in the Hebrew and Christian recognition of the all-seeing eye of God. The application and value of the hints here given may be apparent if the subject of the third eye is studied, and its relation to the spine, and the spinal currents investigated. This third eye is one of the objects of kundalinic vivification, and in the spinal territory there is first the centre at the base of the spine, the home of the sleeping fire. Next we have the triple channel along which that fire will travel in due course of evolution, and finally we find at the summit of the column, and surmounting all, that small organ called the pineal gland, which when vivified causes the third eye to open, and the beauties of the higher, subtler planes to stand revealed. All this physico-psychical occurrence is possible to man owing to certain events which happened to the Heavenly Serpent in the second, or serpent, round. These happenings necessitated the formation and evolution of that peculiar and mysterious family we call the reptilian. These forms of divine life are very intimately connected with the second planetary scheme, being responsive to energy emanating from that scheme, and reaching the earth via the second globe in the second chain. A group of special devas (connected with a particular *open* sound in the planetary Word), work with the reptile evolution. . . .

The *bird kingdom* is specifically allied to the deva evolution. It is the bridging kingdom between the purely deva evolution and two other manifestations of life.

First. Certain groups of devas who desire to pass into the human kingdom, having developed certain faculties, can do so via the bird kingdom, and certain devas who wish to get in communication with human beings can do so via the bird kingdom. This truth is hinted at in the Christian Bible and Christian religious representations by angels or devas being frequently represented as having wings. These cases are not many, as the usual method is for the devas gradually to work themselves towards individualisation through expansive feeling, but in the cases which do occur these devas pass several cycles in the bird kingdom, building in a response to a vibration which will ultimately swing them into the human family. In this way they become accustomed to the use of a gross form without the limitations, and impurities, which the animal kingdom engenders.

Second. Many devas pass out of the group of passive lives in the effort to become manipulating lives via the bird kingdom, and before becoming fairies, elves, gnomes, or other sprites, pass a certain number of cycles in the bird realm.

Why the two above events occur will not be apparent to the casual reader, nor will the true connection between the birds and the devas be accurately realised by the occult student unless he applies himself to the consideration of the "bird or swan out of time and space," and the place that birds play in the mysteries. Herein lies for him the clue. He must remember likewise the fact that every life of every degree, from a god to the most [Page 896] insignificant of the lesser devas, or builders, must at some time or another pass through the human family.

As H. P. B. has pointed out, birds and serpents are closely connected with wisdom, and therefore with the psychic nature of God, of men, and of devas. The study of mythology should reveal certain stages and relationships which will make this matter clearer.

A Treatise on Cosmic Fire, pp. 890-96

THE GREATER AND LESSER BUILDERS 123

2) Elementals of Liquid Matter

• We must centre our attention first upon those lives which constitute the sumtotal of all that is watery, and liquid throughout manifestation, and in dealing with this we must remember that we are concerned with the most occult of investigations, and with matters which are very closely connected with man's evolution.

The many groups of the water devas of the manipulatory class have been roughly grouped by mythological writers, under the terms undines, mermaids, and other expressions, but their diversity is great, and this will be necessarily apparent when it is remembered that the sumtotal of water upon the earth (oceans, seas, rivers, lakes and streams), far exceeds the dry portion, or land, and every drop of moisture is in itself a tiny life, fulfilling its function and running its cycle. The mythic forms above referred to are but those myriad lives built into a form through which an evolutionary deva is seeking expression.

The extreme interest of this subject might be expressed under certain statements which will give the student some idea of the close attention which should, and eventually will be paid to this subject of the deva lives of watery manifestation. As said above, the aggregate of these lives is greater than the aggregate of those lives which form the sum total of solid earth as we understand the term, even though they do not exceed the number of lives which form the gaseous portion of manifestation; this gaseous portion is found in the atmosphere, interpenetrating dense matter, and filling in a large degree the interior caverns of the planet. The microcosmic resemblance to the great Life of the planet is seen in the fact that both forms are but outer sheaths or frameworks, sheltering an inner "vault"; both forms are hollow, both have their negative and positive extremities, their poles, so to speak, and internally much proceeds affecting the outer evolutions.

One of the most occult of the planets, Neptune, presides over the "devas of the waters"; their presiding deva Lord, Varuna, the Raja of the astral plane, being an emanation from that planet. Students will find it of profound interest to study the close interaction therefore between:

124   *THE GREATER AND LESSER BUILDERS*

1. The sixth plane, the astral plane, and the sixth subplane of the physical plane, the liquid subplane.
2. The sixth subplane of each plane in the solar system, and their relation to each other.

Herein will be found one reason why men of a relatively low type of physical body, and having an astral body with some sixth subplane matter in it are responsive to higher things and have a spiritual aspiration. The influence emanating from the sixth subplane of the buddhic plane calls out a reciprocal response from the sixth subplane matter in other bodies, and the sixth principle of buddhi under the Law of Correspondences intensifies that vibration.

A Treatise on Cosmic Fire, pp. 896/8

• The sixth principle, therefore, or the love aspect (the Christ principle), and the sixth plane, are connected; there is an interplay of energy between the fourth cosmic ether, or buddhic energy, and the sixth plane, or astral energy. The devas on both these planes belong essentially to groups over which Neptunian influence presides, hence the astral plane can, and eventually will, directly reflect the buddhic. . . .

All this is revealed to esotericists in the symbology of the circulatory system in man. As the blood system, with its two types of channels (arteries and veins) and its two types of builders (the red and the white corpuscles), is studied from the occult standpoint, much will be ascertained of a revolutionary nature. The laws of the path of outgoing, and of the path of return, with the two groups of deva lives therein concerned, will be apprehended by man. A further hint may here be given. In the physical body of man in connection with the circulatory system, we find, in the three factors—the heart, the arteries, and the veins—the clue to the three types of devas, and also to the systemic triangle which they represent, and further, to the three modes of divine expression. There is a planetary as well as a systemic circulation, and it is carried on through the medium of deva substance everywhere, macrocosmically as well as microcosmically.

A Treatise on Cosmic Fire, pp. 901/2

THE GREATER AND LESSER BUILDERS 125

• A hint in connection with medicine is here to be found; it is occultly true that just as the liquid devas and elementals are closely related to the vegetable kingdom, and both to the plane of the emotions, the logoic liquid body, so the ills of men which affect the circulatory system, the kidneys, the bladder, and the lubrication of the joints, will find a CURE in vegetable constituents and above all in the right adjustment of the emotional nature.

A Treatise on Cosmic Fire, pp. 945

• The devas of water find for themselves the path of service in their great work of nourishing all the vegetable and animal life upon the planet; the goal for them is to enter into that higher group of devas which we call the gaseous or fire devas. These, through the action of their fire upon the waters, produce that sequence of evaporation, condensation, and eventual precipitation which—through its constant activity—nourishes all life upon the earth. Thus again can the psychic laws of love be seen at work in the deva kingdom as in the human; first, the withdrawal or segregation of the unit from the group (called individualisation in man, and evaporation in the water realm). Next, condensation, or the amalgamation of the unit with a newer or higher group, this we call condensation for the devas of the waters, and initiation in man; finally, the sacrifice of the group of human or deva atoms to the good of the whole. So does the law of service and sacrifice govern all the second aspect divine in all its departments great or small. Such is the law. But in the human kingdom, though love is the fulfilling of the law, it is arrived at along the path of pain and sorrow, and every true lover and server of humanity is stretched upon the cross until for them the sixth principle dominates, and the sixth type of matter in their bodies is completely subjected to the higher energy. In the case of the devas, love is the fulfilling of the law without pain or sorrow. It is for them the line of least resistance, for they are the mother aspect, the feminine side of manifestation, and the easy path for them is to give, to nourish, and to tend. Therefore, the devas of the waters pour themselves out in service to the vegetable and animal kingdoms, and in the transmutative fires all that holds them on the sixth subplane will eventually be overcome, and through occult "distillation and evaporation" these devas will eventually form part of the gaseous fiery group and become those fires which are the basis of the divine alchemy.   *A Treatise on Cosmic Fire*, pp. 902/3

3) Elementals of Gaseous Matter

• *The Devas of the Gaseous Subplane*: In dealing with the elementals, or lesser devas, under the manipulatory devas of this extensive group, we are dealing with the devas of fire, and with the fiery essences of substantial nature which can be seen manifesting in myriads of forms. Certain of the subdivisions of this group are known to students, such as:

The Salamanders, or the fiery lives which can be seen by clairvoyants leaping in the flames of a furnace or of a volcano; this group can be subdivided into four groups according to color—red, orange, yellow, and violet—the last of which approximate very closely to the devas of the fourth ether.

The Agnichaitans; this is a term applied to the fiery lives, which are the sumtotal of the plane substance, as seen in the first part of our treatise, and also to the tiny essences which compose the fires of manifestation. As the nature of physical plane electricity is understood and studied, and its true condition realised, the reality of the existence of these agnichaitans will stand revealed.   *A Treatise on Cosmic Fire,* p. 904

• There are, for instance, the denser forms of gaseous life, termed often salamanders, the elementals of the fire. These are directly under the control of the Lord Agni, Lord of the mental plane, and, in this mental age, we have the element of fire entering into the mechanics of living as never before. Eliminate the products which are controlled by heat and you will bring our civilisation to a stop; you will bring all means of transportation to an end and all modes of lighting; you would throw all manufactories into the discard. Basically again, these fiery lives, are found in all that burns, and in the warmth that holds all life formation on earth and causes the flourishing of all living things.
A Treatise on White Magic, p. 389

THE GREATER AND LESSER BUILDERS 127

• These agnichaitans of the third subplane come particularly under the influence of Saturnian energy. They are the great fusers of substance, and it is in connection with them that the transmutation of metals becomes possible. They have a relationship to the mineral kingdom analogous to that which the watery devas have to the vegetable and animal. They are, as will be apparent, connected with the throat centre of a planetary Logos and of a solar Logos, and it is through their activity that the transmission of sound through the air becomes possible. It might surprise students and inventors could they but realise that the present rapid growth of wireless communication everywhere is due to the swinging into contact with the human vibration of a group of fiery deva lives hitherto uncontacted.

A Treatise on Cosmic Fire, p. 905

• This brings us to the point which we are seeking to make anent this third group of the lowest devas. They are very destructive where man is concerned, for they embody the final and therefore powerful vibration of the past system, the conscious activity of dense matter. Hence there is consequently a profound truth in the statement that man is at the "mercy of the elements." Man can physically be burned and destroyed by fire; he is helpless before volcanic action, and cannot protect himself from the ravages of fire, unless in the initial stages of such deva endeavour. The occult importance of the war man wages on the fire devas for instance is very real in connection with the fire department in any city. The time lies far ahead as yet, but it will surely come, when the personnel of these departments will be chosen for their ability to control the agnichaitans when manifesting destructively, and their methods will no longer be that of water (or the calling in of the water devas to neutralise the fire devas), but that of incantation, and a knowledge of the sounds that will swing into action forces which will control the fiery destructive elements.

A Treatise on Cosmic Fire, pp. 639/40

• . . . the Mahachohan is working specifically at this time (in cooperation with the Manu), with the devas of the gaseous subplane;

this is in connection with the destroying work they are to effect by the end of this rootrace, in order to liberate Spirit from constricting forms. Volcanic action therefore may be looked for, demonstrating in unexpected localities, as well as within the sphere of the present earthquake and volcanic zones. Serious disturbance may be looked for in California before the end of the century, and in Alaska likewise. *A Treatise on Cosmic Fire,* p. 907

• Another angle of the Mahachohan's work at this time is connected with *sound,* and therefore with the particular devas whom we are considering. Through the mismanagement of men, and their unbalanced development, the sounds of earth, such as those of the great cities, of the manufactories, and of the implements of war, have brought about a very serious condition among the gaseous devas. This has to be offset in some way and the future efforts of civilisation will be directed towards the spreading of a revolt against the evils of congested living and to the dissemination of an impulse of a widespread nature to seek the country and wide spaces. One of the main interests in the future will be a tendency towards the elimination of noise, owing to the increased sensitiveness of the race. When the energy of water and of the atom is harnessed for the use of man, our present types of factories, our methods of navigation and of transportation, such as steamers and railway apparatus, will be entirely revolutionised. This will have a potent effect not only on man but on the devas. *A Treatise on Cosmic Fire*, pp. 909/10

4) Elementals of the Ethers

• Devas of all kinds and colours are found on the physical etheric levels, but the prevailing hue is violet, hence the term so often employed, the "devas of the shadows." With the coming in of the ceremonial ray of violet, we have the amplification therefore of the violet vibration, always inherent on these levels, and the great opportunity therefore for contact between the two kingdoms. It is in the development of etheric vision (which is a capacity of the physical human eye)

THE GREATER AND LESSER BUILDERS 129

and not in clairvoyance that this mutual apprehension will become possible. With the coming in likewise of this ray will arrive those who belong thereon, with a natural gift of seeing etherically. Children will frequently be born who will see etherically as easily as the average human being sees physically; as conditions of harmony gradually evolve out of the present world chaos, devas and human beings will meet as friends.

As the two planes, astral and physical, merge and blend, and continuity of consciousness is experienced upon the two, it will be difficult for human beings to differentiate at first between devas of the astral plane, and those of the physical. At the beginning of this period of recognition, men will principally contact the violet devas, for those of the higher ranks amongst them are definitely making the attempt to contact the human. These devas of the shadows are of a dark purple on the fourth etheric level, of a lighter purple, much the same colour as violet, on the third etheric level, a light violet on the second, whilst on the atomic subplane they are of a glorious translucent lavender.

Some of the groups of devas to be contacted on the physical plane are as follows:

Four groups of violet devas, associated with the etheric doubles of all that exists on the physical plane. These four are in two divisions, those associated with the building of the etheric doubles, and those out of whose substance these doubles are built.

The green devas of the vegetable kingdom. These exist in two divisions also. They are of high development, and will be contacted principally along the lines of magnetisation. The greater devas of this order preside over the magnetic spots of the earth, guard the solitude of the forests, reserve intact spaces on the planet which are required to be kept inviolate; they defend them from molestation, and with the violet devas are at this time working definitely, though temporarily, under the Lord Maitreya. The Raja Lord of the astral plane, Varuna and his brother Kshiti, have been called to the council chamber of the Hierarchy for specific consultation, and just as the Masters are endeavouring to prepare humanity for service when the World Teacher comes, so these Raja Lords are working along similar lines in connection with the devas. They are arduous in Their work, intense in Their zeal, but much obstructed by man.

The white devas of the air and water who preside over the atmosphere work with certain aspects of electrical phenomena, and control the seas, rivers, and streams. From among them, at a certain stage in their evolution, are gathered the guardian angels of the race when in physical plane incarnation. Each unit of the human family has his guardian deva.

Each group of devas has some specific method of development and some means whereby they evolve and attain a particular goal.

For the *violet* devas the path of attainment lies through feeling, and through educating the race in the perfecting of the physical body in its two departments.

For the *green* devas the path of service is seen in magnetisation, of which the human race knows nothing as yet. Through this power they act as the protectors of the vegetable plant life, and of the sacred spots of the earth; in their work lies the safety of man's body, for from the vegetable kingdom for the remainder of this round comes the nourishment of that body.

For the *white* devas the path of service lies in the guarding of the individuals of the human family, in the care and segregation of types, in the control of the water and air elementals, and much that concerns the fish kingdom.

Thus in the service of humanity in some form or another lies attainment for these physical plane devas. They have much to give and do for humanity, and in time it will be apparent to the human unit what he has to give towards the perfecting of the deva kingdom. A great hastening of their evolution goes forward now coincident with that of the human family.

A Treatise on Cosmic Fire, pp. 911/4

6. MAN–A BUILDER IN MENTAL MATTER

- This Treatise seeks to prove, that in the fourth kingdom the three fires meet:

  a. Fire by friction, or the negative Brahma Aspect, the third Aspect.
  b. Solar Fire, or the positive negative Vishnu Aspect, the second Aspect.
  c. Electric Fire, or the positive Shiva Aspect, the first Aspect.

Man in the three worlds, consciously or unconsciously, recapitulates the logoic process, and becomes a creator, working in substance through the factor of his positive energy. He wills, he thinks, he speaks, and thought-forms eventuate. Atomic substance is attracted to the enunciator. The tiny lives which compose that substance are forced (through the energy of the thinker), into forms, which are themselves active, vitalised and powerful. What man builds is either a beneficent or a maleficent creation according to the underlying desire, motive, or purpose.

It is essential that we endeavour to make practical what is here to be imparted, as it is useless for man to study the groups of lesser builders, their functions and their names, unless he realises that with many of them he has an intimate connection, being himself one of the great builders, and a creator within the planetary scheme. Men should remember that through the power of their thoughts and their spoken words they definitely produce effects upon other human beings functioning on the three planes of human evolution and upon the entire animal kingdom. The separative and maleficent thoughts of man are largely responsible for the savage nature of wild beasts, and the destructive quality of some of nature's processes, including certain phenomena, such as plague and famine.

It is of no value to man to know the names of some of the "army of the voice" unless he comprehends his relationship to that army, unless he apprehends the responsibility which is his to be a beneficent creator, working under the law of love, and not impelled to the creative act through selfish desire, or uncontrolled activity.

A Treatise on Cosmic Fire, pp. 888/9

• Humanity is intended to be the medium wherein certain activities can be instituted. It is in reality the brain of the planetary Deity, its many units being analogous to the brain cells in the human apparatus. Just as the human brain, made up of an infinite number of sentient responsive cells, can be suitably impressed when quiescence has been achieved, and can become the medium of expression for the plans and purposes of the soul, transmitting its ideas via the mind, so the planetary Deity, working under the inspiration of the Universal Mind, can impress humanity with the purposes of God and produce consequent effects in the world of phenomena.

A Treatise on White Magic, p. 527

• Through speech a thought is evoked and becomes present; it is brought out of abstraction and out of a nebulous condition and materialised upon the physical plane, producing (could we but see it) something very definite on etheric levels. Objective manifestation is produced, for "Things are that which the Word makes them in naming them." Speech is literally a great magical force, and the adepts or white magicians, through knowledge of the forces and power of silence and of speech, can produce effects upon the physical plane. As we well know, there is a branch of magical work which consists in the utilisation of this knowledge in the form of Words of Power and of those mantrams and formulae which set in motion the hidden energies of nature and call the devas to their work.

A Treatise on Cosmic Fire, p. 981

• It might be of value here if students realised that every good speaker is doing a most occult work. A good lecturer (for instance) is one who is doing work that is analogous on a small scale to that done by the solar Logos. What did He do? He thought, He built, He vitalised. A lecturer, therefore, segregates the material with which he is going to build his lecture and which he is going to vitalise. Out of all

THE GREATER AND LESSER BUILDERS 133

the thought matter of the world he gathers together the substance which he individually seeks to use. Next he copies the work of the second Logos in wisely building it into form. He constructs the form, and then when it is constructed, he finishes up by playing the part of the first Person of the Trinity putting his Spirit, vitality and force into it so that it is a vibrant, living manifestation. When a lecturer or speaker of any kind can accomplish that, he can always hold his audience and his audience will always learn from him; they will recognise that which the thought form is intended to convey.

A Treatise on Cosmic Fire, p. 979

• The devas who are the sumtotal of the energy of substance itself care not what form they build. They are irresponsibly responsive to energy currents, and theirs is not the problem of dealing with sources of energy. Therefore, the place of man in the cosmic plan becomes more vital and apparent when it is realised that one of his main responsibilities is the direction of energy currents from the mental plane, and the creation of that which is desired on higher levels. Men, as a whole, are undergoing evolutionary development in order that they may become conscious creators in matter. This involves

A realisation of the archetypal plan,

An understanding of the laws governing the building processes of nature,

A conscious process of willing creation, so that man co-operates with the ideal, works under law, and produces that which is in line with the planetary plan, and which tends to further the best interests of the race,

A comprehension as to the nature of energy, and an ability to direct energy currents, to disintegrate (or withdraw energy from) all forms in the three worlds,

An appreciation of the nature of the devas, their constitution and place as builders, and of the words and sounds whereby they are directed and controlled.

When the energy currents of the human family are directed from egoic levels only, when desire is transmuted, and the fifth principle

THE GREATER AND LESSER BUILDERS

CHART VIII
THE EGOIC LOTUS AND THE CENTRES
COSMIC PHYSICAL PLANE

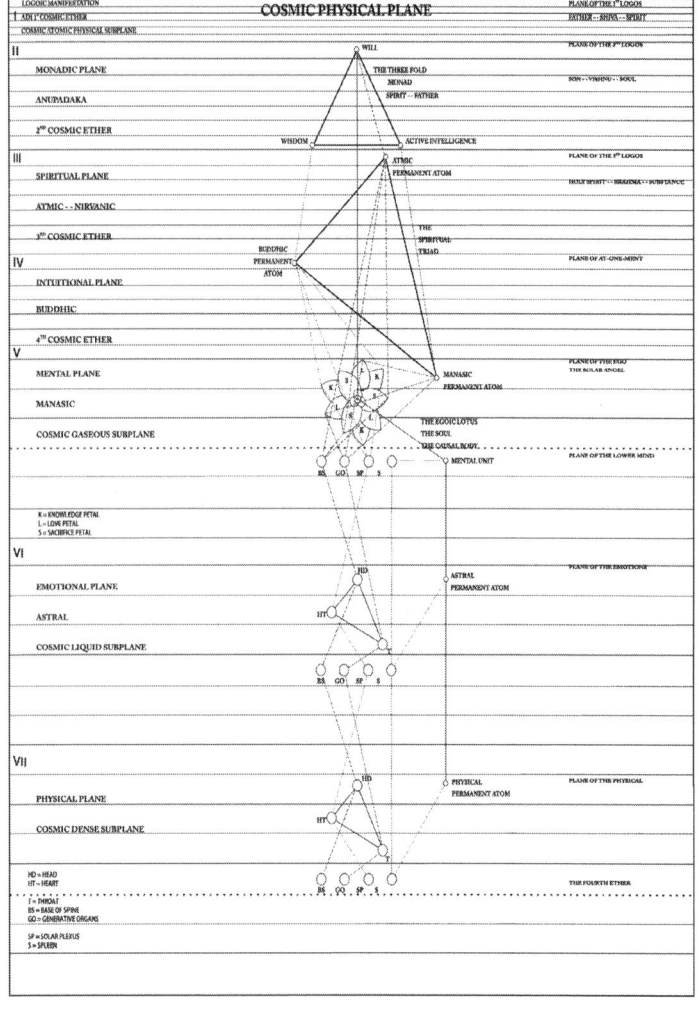

A Treatise on Cosmic Fire, p. 817

awakened and finally illuminated by the sixth, then and only then will the strength of the impulse emanating from lower levels die out and the "Dweller on the Threshold" (who now haunts the human family) likewise die. In other words, when the dense physical body of the planetary Logos (composed of matter of the three worlds of human endeavour) is completely purified and vitalised by the force of the life flowing from etheric levels, and when all His centres (formed of human units) are fully awakened, then will those centres be channels for pure force, and such an entity as the "Dweller" be an impossibility.

A Treatise on Cosmic Fire, pp. 951/3
See also: 'Man as a Creator in Mental Matter',
A Treatise on Cosmic Fire, pp. 947/63
and 'Man and the Fire Spirits or Builders', pp. 963/8

7. THE HUMAN SOUL AND THE PERMANENT ATOMS

•THE SOLAR ANGELS-THE AGNISHVATTAS

We start here upon the consideration of the Agnishvattas, or the Fire devas of the mental plane, and are thus launched upon the most stupendous subject in connection with our planetary evolution; it is the one having the most occult significance for man, for these solar Angels concern his own essential nature, and are also the creative power by which he works. For all practical purposes, and for the elucidation of the spiritual evolution of man, this immediate section is of the greatest interest and importance; it should be one of the most widely studied sections of this treatise. Man is ever profoundly interested in himself, and before he can duly develop must comprehend *scientifically* the laws of his own nature, and the constitution of his own "mode of expression." He must realise likewise somewhat of the inter-relation of the three fires in order that he may himself at some future date "blaze forth."

The question of these Fire Dhyanis and their relation to man is a most profound mystery, and the entire matter is so clothed in intricate legends that students are apt to despair of ever arriving at the desired, and necessary clarity of thought. Not yet will it be possible entirely to dispel the clouds which veil the central mystery, but perhaps, by due tabulation and synthesis, and by a cautious amplification of the data already imparted, the thoughts of the wise student may become somewhat less confused.

A Treatise on Cosmic Fire, pp. 679/80

• *The Solar Angels and the Fifth Principle.*
 . . . Where man is concerned these solar Angels, the Agnishvattas, produce the union of the spiritual Triad, or divine Self, and the Quaternary, or lower self. Where the Logos is concerned, whether solar or planetary, they produce conditions whereby the etheric, and the dense physical become a unit.

They represent a peculiar type of electrical force; their work is to blend and fuse, and above all else they are the "transmuting fires" of the system, and are those agents who pass the life of God through their bodies of flame as it descends from the higher into the lower, and again as it ascends from the lower into the higher. They are connected in their highest groups with that portion of the logoic head centre which corresponds to the heart, and here is the clue to the mystery of kama-manas. The kamic angels are vitalised from the "heart" centre and the manasic angels from the logoic head centre, *via the point within that centre connected with the heart.* These two dominating groups are the sumtotal of kama-manas in all its manifestations. The solar angels exist in three groups, all of which are concerned with the self-consciousness aspect, all of which are energised and connected with the fifth spirilla of the logoic permanent atom, and all of which work as a unit.

One group, the highest, is connected with the logoic head centre, whether solar or planetary. They work with the manasic permanent atoms and embody the will-to-be in dense physical incarnation. Their power is felt on the atomic subplane and on the second; they are the substance and the life of those planes. Another group is connected definitely with the causal bodies of all Egos and are of prime importance in this solar system. They come from the heart centre, and express that force. The third group, corresponding to the throat centre, show forth their power on the fourth subplane through the mental units. They are the sumtotal of the power of the Ego to see, to hear and to speak (or sound) in the strictly occult sense.

A Treatise on Cosmic Fire, pp. 698/9

• On Individualisation.

(a.) The Work of the Solar Angels. Let us briefly consider the general construction of the body of the Ego enumerating its component parts and bearing in mind that the form is ever prepared prior to occupancy. From the study of this body, we can get some idea of, and some light upon macrocosmic Individualisation.

The causal body, called sometimes (though inaccurately) the "karana sarira," has its place on the third subplane of the mental plane, the lowest abstract plane, and the one whereon the Ray of the third Logos provides the necessary "light for construction." (This is because each subplane comes specially under the influence of its Number, Name, or Lord.) When the hour strikes and the vehicles for buddhi are to be co-ordinated certain great Beings, Lords of the Flame, or Manasadevas, through driving external force, come in conjunction with the material of that subplane, and vitalise it with Their Own energy. They form a new and positive impulse which co-ordinates the material of the plane and produces a temporary balancing of forces. Hence the meaning of the "white," or transparent condition of the new causal body. It remains with the new-born ego first to upset the equilibrium, and then to regain it, at the close of the process, producing a radiant form, full of primal colours.

At the coming in of the Manasadevas to produce self-consciousness and to bring about the incarnation of the divine Egos, four things occur on that plane. If the student adds to these four those which have been already imparted in various occult books anent the effect of individualisation on animal man and his appearance as a self-conscious identity on the physical plane, a working hypothesis is provided whereby man can scientifically undertake his own unfoldment. These four are given in the order of their appearance in time and space:

First. There appear upon the third subplane of the mental plane certain vibratory impulses—nine in number—corresponding to the fivefold vibration of these Manasadevas in conjunction with the fourfold vibration set up from below and inherent in the matter of this subplane, the fifth from the lower standpoint. This produces "the ninefold egoic lotus," which is at this stage tightly closed, the nine petals folded one upon the other. They are vibrant, and scintillating "light" but not of excessive brightness. These "lotus buds" are in groups, according to the influence of the particular ones of the fivefold Dhyanis Who are acting upon it and Who form it out of Their own substance, colouring it faintly with the "fire of manas.

Second. There appears a triangle on the mental plane, produced by manasic activity, and this triangle of fire begins slowly to circulate between the manasic permanent atom, and a point at the centre of the egoic lotus, and thence to the mental unit, which has appeared upon the fourth subplane through innate instinct approximating mentality. This triangle of fire, which is formed of pure electrical manasic force, waxes ever brighter until it produces an answering vibration from both the lower and the higher. This triangle is the nucleus of the antahkarana. The work of the highly evolved man is to reduce this triangle to a unity, and by means of high aspiration (which is simply transmuted desire affecting mental matter) turn it into the Path and thus reproduce in a higher synthetic form the earlier "path" along which the descending Spirit came to take possession of its vehicle, the causal body, and from thence again work through the lower personal self.

Third. At a certain stage of vibratory activity, the work of the Lords of the Flame having produced a body or form and a vibration calling for response, there occurs a practically simultaneous happening.

A downflow of buddhi takes place along the line of the manasic triangle until it reaches a point at the very centre of the lotus. There, by the power of its own vibration, it causes a change in the appearance of the lotus. At the very heart of the lotus, three more petals appear which close in on the central flame, covering it closely, and remaining closed until the time comes for the revelation of the "jewel in the Lotus." The egoic lotus is now composed of twelve petals, nine of these appear at this stage in bud form and three are completely hidden and mysterious.

At the same time, the three permanent atoms are enclosed within the lotus, and are seen by the clairvoyant as three points of light in the lower part of the bud, beneath the central portion. They form at this stage a dimly burning triangle. The causal body, though only in an embryonic condition, is now ready for full activity as the eons slip away, and is complete in all its threefold nature. *The matter aspect,* which concerns the material form of the man in the three worlds, or his active intelligent personal self can be developed and controlled through the medium of the mental unit, the astral permanent atom and the physical permanent atom. *The Spirit aspect* lies concealed at the heart of the lotus, in due course of time to stand revealed when the manasadevas have done their work. The will that persists forever is there. *The consciousness aspect* embodying the love-wisdom of the divine Ego as it reveals itself by the means of mind is predominantly there, and in the nine petals and their vibratory capacity lies hid all opportunity, all innate capacity to progress, and all the ability to function as a self-conscious unit, that entity we call Man. Mahadeva sits at the heart, Surya or Vishnu reveals Him in His essence as the Wisdom of Love and the Love of Wisdom, and Brahma, the Creative Logos makes that revelation possible. The Father in Heaven is to be revealed through the Christ, the Son, by the method of incarnation made possible through the work of the Holy Spirit. All this has been brought about by the sacrifice and instrumentality of certain cosmic entities who "offer Themselves" up in order that Man may be. From their very essence, they give out that which is needed to produce the individualising principle, and that which we call "self-consciousness," and thus enable the divine Spirit to enter into fuller life by means of limitation by form, by means of the lessons garnered through a long pilgrimage, and through the "assimilation of manifold existences."

The *fourth* point to be noted is that when these three events have occurred, the light or fire that circulates along the manasic triangle is withdrawn to the centre of the lotus, and this "prototype" of the future antahkarana, if so it may be expressed, disappears. The threefold energy of the petals, the atoms and the "jewel" is now centralised, because impulse must now be generated which will produce a downflow of energy from the newly made causal vehicle into the three worlds of human endeavour.

We have dealt with the method of individualisation through the coming in of the Lords of the Flame because it is the prime method in this solar system; whatever methods may be pursued in the varying schemes and chains, this—at the middle stage—is the universal rule. Karmic conditions having to do with a planetary Logos may effect modifications, and bring into action manasadevas whose activity may not be the same in working detail, but the results are ever similar, and the divine Egos in their causal bodies have analogous instruments to work through....

A *final* point which is of profound significance is that *the Agnishvattas construct the petals out of Their Own substance, which is substance energised by the principle of "I-ness," or ahamkara. They proceed to energise the permanent atoms with Their own positive force, so as to bring the fifth spirilla in due course of time into full activity and usefulness.* All possibility, all hopefulness and optimism, and all future success lies hid in these two points.

As we have seen, the work of the Agnishvattas on the mental plane resulted in a downflow of force or energy from the Monad (or Spirit) and this, in conjunction with the energy of the lower quaternary produced the appearance of the body of the Ego on the mental plane. In ordinary electric light, we have a faint illustration of the thought I seek to convey. By the approximation of the two polarities, light is created. By an analogous type of electrical phenomena, the light of the Monad shines forth, but we have to extend the idea to the subtler planes, and deal with seven types of force or energy in connection with the one polarity and with four in connection with the other. A scientific formula for the process of individualisation conveys this dual approximation with its differing types of energy in one symbol and a number, but it cannot here be revealed.

The Manasadevas are themselves energised by force from the cosmic mental plane—a force which has been in operation ever since the individualisation of the solar Logos in kalpas far distant. They, in Their corporate nature, embody the will or purpose of the Logos, and are the cosmic "prototypes" of our solar Angels. The solar Angels on the mental plane of the system embody as much of that will and purpose as the Logos can work through in one single incarnation and as They, in Their groups, can develop. They work, therefore, through egoic groups and primarily, after individualisation, upon the mental units of the separated identities who go to the constituency of the groups. This is Their secondary work. Their work in part might be described as follows:

Primarily, they bring about the union of the divine Ego and the lower personal self. This has been dealt with.

Secondly, they work through the mental units, impressing upon the atom that portion, microscopic as it may be, of the logoic purpose which the individual can work out on the physical plane. At first their influence is unconsciously assimilated, and the man responds to the plan blindly and ignorantly. Later, as evolution proceeds, their work is recognised by the man in a conscious co-operation with the plan of evolution. After the third initiation, the will or purpose aspect predominates.

It might here be noted that it is the positive force of the Manasadevas that produces initiation. Their function is embodied by the Hierophant. He, seeing before Him the vehicle for buddhi, passes the voltage from the higher planes through His body, and by means of the Rod (charged with positive manasic force) transmits this higher manasic energy to the initiate so that he is enabled to know consciously and to recognise the plan for his group-centre through the immensely increased stimulation. This force descends from the manasic permanent atom via the antahkarana and is directed to whichever centre the Hierophant—under the Law—sees should be stimulated. He stabilises the force, and regulates its flow as it circulates throughout the egoic Lotus, so that when the work of unfoldment is accomplished the sixth principle at the Heart of the Lotus can stand revealed. After each initiation the Lotus is more unfolded and light from the centre begins to blaze forth—a light or fire which ultimately burns through the three enshrining petals, and permits the full inner glory to be seen, and the electric fire of spirit to be manifested. As this is brought about on the second subplane of the mental plane (whereon the egoic lotus is now situated) a corresponding stimulation takes place in the dense substance which forms the petals or wheels of the centres on the astral and etheric levels.

A Treatise on Cosmic Fire, pp. 707/14

Permanent Atoms

• . . . it should be remembered that all the planes of our system, viewing them as deva substance, form the spirillae in the physical permanent atom of the solar Logos. This has earlier been pointed out, but needs re-emphasising here. All consciousness, all memory, all faculty is stored up in the permanent atoms, and we are consequently dealing here with that consciousness; the student should nevertheless bear in mind that it is on the atomic subplanes that the logoic consciousness (remote as even that may be from the Reality) centres itself. This permanent atom of the solar system, which holds the same relation to the logoic physical body as the human permanent atom does to that of a man, is a recipient of force, and is, therefore, receptive to force emanations from another extra-systemic source. Some idea of the illusory character of manifestation, both human and logoic, may be gathered from the relation of the permanent atoms to the rest of the structure. Apart from the permanent atom, the human physical body does not exist.

Again, forms differ as do kingdoms according to the nature of the force flowing through them. In the animal kingdom that which corresponds to the permanent atom responds to force of an involutionary character, emanating from a particular group. The human permanent atom responds to force emanating from a group on the evolutionary arc and the Ray of a particular planetary Logos in Whose body a human Monad has a definite place.

A Treatise on Cosmic Fire, pp. 693/4

• *The purpose of the permanent atoms* - The three permanent atoms are in themselves centres of force, or those aspects of the personality which hold hid the fires of substance, or of objectivity; it cannot be too strongly pointed out at this juncture that, in considering the threefold man in the three worlds, we are dealing with substance which (in connection with logoic manifestation) is considered the dense physical. Surrounding these three atoms is the causal sheath, answering the following purposes:

It separates one unit of egoic consciousness from another unit of consciousness, yet is itself part of the gaseous body (the fifth cosmic physical subplane) in the physical body of the planetary Logos, Who is the central life of any particular group of Monads. This fact has been little appreciated, and merits careful consideration. It holds hid spiritual potentialities in its inherent ability to respond to the higher vibration; from the moment of individualisation till it is discarded at initiation, the life within steadily develops these potentialities, and produces certain definite results by the utilisation of the three permanent atoms. It gradually vivifies and awakens them until, on the three planes, the central life has an adequate point of contact which can originate the necessitated vibration in the matter of the plane.

The permanent atoms on each plane serve a fourfold purpose as regards the central or egoic life:

They are the distributors of a certain type of force.

They are the conservers of faculty or ability to respond to a particular vibration.

They are the assimilators of experience and the transmuters of that experience into quality. This is the direct result of the work of the egoic Ray as it plays upon the atom.

They hold hid the memory of the unit of consciousness. When fully vibrant they are the raison d'être for the continuity of the consciousness of the man functioning in the causal body. This distinction must be carefully made.

A Treatise on Cosmic Fire, pp. 507/8

• The permanent atoms are enclosed within the periphery of the causal body, yet that relatively permanent body is built and enlarged, expanded and wrought into a central receiving and transmitting station (using inadequate words to convey an occult idea) by the direct action of the centres, and of *the centres above all*. Just as it was spiritual force, or the will aspect, that built the solar system, so it is the same force in the man that builds the causal body. By the bringing together of spirit and matter (Father-Mother) in the macrocosm, and their union through the action of the will, the objective solar system, or the Son, was produced—that Son of desire, Whose characteristic is love, and Whose nature is buddhi or spiritual wisdom. By the bringing together (in microcosm) of Spirit and matter, and their coherence by means of force (or the spiritual will) that objective system, the causal body, is being produced; it is the product of transmuted desire, whose characteristic (when fully demonstrated) will be love, the expression eventually on the physical plane of buddhi. The causal body is but the sheath of the Ego. The solar system is but the sheath of the Son. In both the greater and the lesser systems, force centres exist which are productive of objectivity. The centres in the human being are reflections in the three worlds of those higher force centres.

A Treatise on Cosmic Fire, pp. 178/9

• The appearance, and the final disappearance, of any manifested Life is intimately concerned with the possession, the evolutionary development, and the final disintegration, of the permanent atom. Permanent atoms, as the term is usually understood, are the property of those lives only who have achieved self-consciousness, or individuality, and therefore relative permanence in time and space. The permanent atom may be viewed as the focal point of manifestation on any particular plane. It serves, if I may use so peculiar a term, as the anchor for any particular individual in any particular sphere, and this is true of the three great groups of self-conscious Lives:

a. The incarnating Jivas, or human beings,
b. The planetary Logoi,
c. The solar Logos.

We must remember here that all the atomic subplanes of the seven planes form the seven spirillae of the logoic permanent atom, for this has a close bearing upon the subject under consideration.

The units, therefore, in the three lower kingdoms possess no permanent atoms but contribute to the formation of those atoms in the higher kingdoms. Certain wide generalisations might here be made, though too literal or too identified an interpretation should not be put upon them.

First, it might be said that the lowest or *mineral kingdom* provides that vital something which is the essence of the physical permanent atom of the human being. It provides that energy which is the negative basis for the positive inflow which can be seen pouring in through the upper depression of the physical permanent atom.

Secondly, the *vegetable kingdom* similarly provides the negative energy for the astral permanent atom of a man, and thirdly, the *animal kingdom* provides the negative force which when energised by the positive is seen as the mental unit. This energy which is contributed by the three lower kingdoms is formed of the very highest vibration of which that kingdom is capable, and serves as a link between man and his various sheaths, all of which are allied to one or other of the lower kingdoms.

> *a.* The mental body.......mental unit....................animal kingdom.
> *b.* The astral body.........astral permanent atom..........vegetable kingdom.
> *c.* The physical body......physical permanent atom......mineral kingdom.

In man these three types of energy are brought together, and synthesised, and when perfection of the personality is reached, and the vehicles aligned, we have:

> *a.* The energy of the mental unit..........................positive.
> *b.* The energy of the astral permanent atom........equilibrised.
> *c.* The energy of the physical permanent atom.....negative.

Man is then closely linked with the three lower kingdoms by the best that they can provide, and they have literally given him his permanent atoms, and enabled him to manifest through their activity.

A Treatise on Cosmic Fire, pp. 1133/5

8. HUMAN AND DEVA EVOLUTION COMPARED

• The fourth Creative Hierarchy, that of the human Monads, has to learn to vibrate positively, but the devas proceed along the line of least resistance; they remain negative, taking the line of acquiescence, of falling in with the law. Only the human Monads, and only in the three worlds, follow the positive line, and by resistance, struggle, battle and strife learn the lesson of *divine* acquiescence. Yet, owing to the increase of friction through that very struggle, they progress with a relatively greater rapidity than the devas. They have need to do this, for they have lost ground to make up.

A Treatise on Cosmic Fire, p. 575

• In the three worlds, we have the parallel evolutions—deva and human in their many varying grades—the human naturally concerning us the most intimately, though the two evolve through interaction with each other. In the higher four worlds, we have this duality viewed as a unity, and the aspect of the synthetic evolution of the Heavenly Men is the one considered. It would interest us much could we but understand a little of the point of view of those great devas Who cooperate intelligently in the plan of evolution. They have Their own method of expressing these ideas, the medium being colour which can be heard, and sound which can be seen. Man reverses the process and sees colors and hears sounds. A hint lies here as to the necessity for symbols, for they are signs which convey cosmic truths, and instruction, and can be comprehended alike by the evolved of both evolutions. It should be borne in mind, as earlier pointed out, that:

a. Man is demonstrating the aspects of divinity. The devas are demonstrating the attributes of divinity.

b. Man is evolving the inner vision and must learn to see. The devas are evolving the inner hearing and must learn to hear.

c. Both are as yet imperfect, and an imperfect world is the result.

d. Man is evolving by means of contact and experience. He expands. The devas evolve by means of the lessening of contact. Limitation is the law for them.

e. Man aims at self-control. Devas must develop by being controlled.

f. Man is innately Love,—the Force which produces coherency. The devas are innately intelligence,—the force which produces activity.

g. The third type of force, that of Will, the balancing equilibrium of electrical phenomena, has to play equally upon and through both evolutions, but in the one it demonstrates as self-consciousness, and in the other as constructive vibration.

In the Heavenly Man these two great aspects of divinity are equally blended, and in the course of the mahamanvantara the imperfect Gods become perfect. These broad and general distinctions are pointed out as they throw light upon the relationship of Man to the devas. *A Treatise on Cosmic Fire,* pp. 666/7

• More and more, during the next fifteen years, will men receive definite teaching, often subconsciously, from devas to whom they are linked. This will be done telepathically at first. Doctors today get much information from certain devas. There are two great devas belonging to the green group on mental levels who assist in this work, and some physicians get much knowledge subjectively from a violet deva working on the atomic subplane of the physical plane, aided by a deva of the causal level who works with, or through, their egos. As men learn to sense and recognise these devas, more and more teaching will be given. They teach in three ways:

    a. By means of intuitional telepathy.
    b. Through demonstration of colour, proving the accomplishment of certain things in this way.
    c. By definite musical sounds, which will cause vibrations in the ethers, which in their turn will produce forms.

The ether will eventually appear to the enhanced vision of humanity to have more substance than it now has, and as etheric vision increases, the ethers will be recognised as being strictly physical plane matter. Therefore when in sickness men shall call a deva, when that deva can destroy diseased tissue by sounding a note that will result in the elimination of the corrupt tissue, when by the presence caused by the vibration new tissue is visibly built in, then the presence of these devas will be generally acknowledged and their power will be utilised.
Esoteric Psychology I, pp. 125/6

• It is in the realisation of these facts anent deva substance, the power of sound, the law of vibration, and the ability to produce forms in conformity with law, that the true magician can be seen.
A Treatise on Cosmic Fire, p. 930

• [Published 1936 – Predictions About the Devas) It is possible at this time to foretell certain events which will come to pass during the next one hundred years.

First, in about ten years time the first ether, with all that is composed of that matter, will be recognised scientific fact, and the scientists who work intuitively will come to recognise the devas of that plane. People coming into incarnation on this seventh ray will have the eyes that see, and the purple devas and the lesser devas of the etheric body will be visioned by them.

Secondly, when He Whom both angels and men await, will approach near unto this physical plane, He will bring with Him not only some of the Great Ones and the Masters, but some of the Devas who stand to the deva evolution as the Masters to the human. Forget not that the human evolution is but one of many, and that this is a period of crisis among the devas likewise. The call has gone forth for them to approach humanity, and with their heightened vibration and superior knowledge unite their forces with those of humanity, for the progression of the two evolutions. They have much to impart anent colour and sound, and their effect upon the etheric bodies of men and animals. When that which they have to give is apprehended by the race,

THE GREATER AND LESSER BUILDERS 149

physical ills will be nullified and attention will be centred upon the infirmities of the astral or emotional body.

These violet devas of the four ethers form, as you may imagine, four great groups with seven subsidiary divisions. These four groups work with the four types of men now in incarnation, for it is a statement of fact that at no time in this round are more than four types of men in incarnation at any one time. Four rays dominate at any given period, with one in excess of the other three. I mean by this, that only four rays are in physical incarnation; for on the plane of the soul all seven types are of course found. This idea is brought out in the four castes in India, and you will find that these four are found universally. The four groups of devas are a band of servers to the Lord, and their special work is to contact men and to give them definite and experimental teaching.

They will instruct in the effect of colour in the healing of disease, especially the effect of the violet light in the lessening of human ills and in the cure of those physical plane sicknesses which have their origin in the etheric body or double.

They will teach men to see etherically, by heightening human vibration by action of their own.

They will demonstrate to the materialistic thinkers of the world the fact that the superconscious states exist—not the superhuman only—and will also make clear the hitherto unrecognised fact that other beings, besides the human, have their habitat on earth.

They will also teach the sounding of the tones that correspond to the gradations of violet, and through that sounding enable man to utilise the ethers, as he now utilises physical plane matter for his various needs.

They will enable human beings so to control the ethers that weight will be for them transmuted, and motion will be intensified, becoming more rapid, more gliding, noiseless, and therefore less tiring. In the human control of the etheric levels lie the lessening of fatigue, rapidity of transit, and the ability to transcend time. Until this prophecy is a fact in consciousness, its meaning is obscure.

They will also teach men how rightly to nourish the body and to draw from the surrounding ethers the requisite food. Man will in the future concentrate more on the sound condition of the etheric body, and the functioning of the dense physical body will become practically automatic.

They will enable human beings, as a race, and not as individuals, to expand their consciousness so that it will embrace the superphysical. Forget not the important fact that in the accomplishment of this the web that divides the physical plane from the astral plane will be discovered by the scientists, and its purpose will eventually be acknowledged. With that discovery will come the power to penetrate the web, and so link up consciously with the astral body. Another material unification will have been accomplished.

Esoteric Psychology I, pp. 123/5

X. CELESTIAL DYNAMICS

1. CONSTELLATIONS

• I should like at this point to make clear the distinction between a constellation and a solar system, according to the esoteric teaching, even though the modern scientist may not agree.

A *solar system* consists of a sun as the central focal point, with its series of attendant planets, which are held in magnetic rapport in their orbits around that sun.

A *constellation* consists of two or more solar systems or series of suns with their attendant planets. These systems are held together as a coherent whole by the powerful interrelation of the suns, whose magnetic rapport is so balanced that occultly "they tread the Path together within the radius of each other's power;" they preserve their relative distances, and vitalise their planets, but at the same time they preserve an equality of balance and of influence. In a few rare cases this balance is disturbed, and there is a waxing or a waning of influence and of magnetic power. This condition is governed by a cosmic law of rhythm so obscure as to be incomprehensible at this time.

An illustration of this waxing and waning of influence and of radiance (synonymous terms in occultism) on a large scale can be seen today in the constellation Gemini, wherein one of the twins is increasing in brilliance and power, and the other is decreasing. But this is a somewhat unique example, esoterically.

Esoteric Psychology I, pp. 152/3

• Students would also do well to remember that the twelve constellations which constitute our particular zodiac are themselves the recipients of many streams of energy coming to them from many sources. These blend and fuse with the energy of any particular constellation and—transmuted and "occultly refined"—eventually find their way into our solar system.

Esoteric Astrology, pp. 13/4

• It might be added in addition that the signs of the zodiac are concerned primarily with the life expression of the Heavenly Man (as far as our planet is concerned) and therefore with the destiny and life of the planetary Logos. They are also concerned with the great *man of the heavens,* the solar Logos. I refer in this last instance to their effect as it makes itself felt in the solar system as a whole and with this effect there are few astrologers at this time fit to deal. I would remind you that to the l*ives who* inform these great constellations and *whose* radiation—dynamic and magnetic—reaches our Earth, this effect is incidental and unnoticed. The primary effect that *they* have is upon our planetary Logos and this effect reaches us through Him, pouring through that great planetary centre to which we have given the name of Shamballa. It is, therefore, capable of evoking the major response from the monads, and these monads express themselves through the kingdom of souls and through the human kingdom; it consequently expresses itself through the Hierarchy and through humanity as a whole. This is a point of real importance and should be noted and connected with all the teaching you have had upon this most interesting theme of the three major planetary centres. It is the work of the zodiacal influences to evoke the emergence of the will aspect of the Heavenly Man and of all monads, souls and personalities who constitute the planetary body of expression. This statement means but little to you today but it will mean much to those students who, in a few decades, will study what I am here saying. Properly understood, it accounts for much that is happening in the world at this time.

Esoteric Astrology, pp. 21/2

• As we ponder and think and as we correlate the various aspects of the teaching, we shall find three propositions emerging which govern the inflow of life to the planet and to the individual man. These have been laid down earlier in *A Treatise on the Seven Rays* but it might profit us to state them here:

Proposition One—Every ray life is an expression of a solar life and every planet is therefore:

CELESTIAL DYNAMICS

1. Linked with every other planetary life.
2. Animated by energy pouring into it from the seven solar systems, of which ours is one.
3. Actuated by three streams of force:
    a. Coming from solar systems other than our own.
    b. Our own solar system.
    c. Our own planetary life.

Proposition Two—Each one of the ray lives is the recipient and the custodian of energies coming from
   1. The seven solar systems.
   2. The twelve constellations.

Proposition Three—It is the quality of a ray life—manifesting in time and space—which determines the phenomenal appearance.

Before we penetrate further into the consideration of our theme, I would like to emphasise two points:

First of all, that we are considering esoteric influences and not astrology, per se. *Our subject is the seven rays and their relationship to the zodiacal constellations* or—in other words—the interaction of the seven great Lives which inform our solar system with the twelve constellations which compose our zodiac.

Secondly, that we have necessarily to study these energies and their interplay from the angle of their effect upon the planet, and incidentally, their effect upon the forms in the various kingdoms of nature and particularly in connection with the fourth kingdom, the human, and with individual man—average man, the disciple and the initiate. . . .

I have pointed out that these energies fall into three groups:

1. Those coming from certain great constellations which are to be found active in relation to our solar system and which, from the most ancient days, have always been related in myth and legend to our system. To these constellations, ours is related in a peculiar way.
2. Those coming from the twelve zodiacal constellations. These are

recognised as having a definite effect upon our system and our planetary life.

3. Those coming from the planets found within the periphery of the Sun's sphere of influence.

From a certain point of view, one can generalise largely and say that these are the correspondences in the solar system to the three great centres of force which produce and control manifestation and evolutionary progress in the human being:

1. The great exterior, yet controlling, constellations are analogous to that centre of force which we call the Monad and to its universal *will-to-power* which is distinctive of the first divine aspect.
2. The twelve constellations might be regarded as embodying the soul aspect and, for the present, their effect upon the individual must be regarded and should be studied in terms of consciousness and of the development of the life of the soul. This is in essence the *will-to-love*.
3. The planets, twelve in number (seven sacred planets and five non-sacred), are effective (using the word in a technical sense) in relation to the external life, environment and circumstances of the individual. Their force contacts should be interpreted largely in terms of the human personality, the third divine aspect. They thus exemplify the *will-to-know*.

Esoteric Astrology, pp. 26/9
See also: 'The Seven Creative Builders, the Seven Rays',
Esoteric Psychology I, pp. 149/57

• Taurus is regarded as the central group of the Milky Way.
Esoteric Astrology, p. 679

• "The Pleiades are the central group of the system of sidereal astronomy.

    a. They are found in the neck of the Bull, the constellation Taurus.

    b. They are therefore in the Milky Way.

c. They are thus considered (Alcyone, in particular) as the central point around which our universe of fixed stars revolves."
S.D. II. 581-582. 3rd ed. [1st ed: 551]
Esoteric Astrology, p. 657

• Lying behind all the many interlocking triangles in our solar system and conditioning them to a very large extent (though today more potentially than expressively) are three energies coming from three major constellations. They are the emanations from the Great Bear, from the Pleiades and from Sirius. It might be pointed out that:

1. The energies coming from the Great Bear are related to the will or purpose of the solar Logos and are to this great Being what the monad is to man. This is a deep mystery and one which even the highest initiate cannot yet grasp. Its sevenfold unified energies pass through *Shamballa*.
2. The energies coming from the sun, Sirius, are related to the love-wisdom aspect or to the attractive power of the solar Logos, to the soul of that Great Being. This cosmic soul energy is related to the Hierarchy. You have been told that the great White Lodge on Sirius finds its reflection and a mode of spiritual service and outlet in the great White Lodge of our planet, the *Hierarchy*.
3. The energies coming from the Pleiades, an aggregation of seven energies, are connected with the active intelligent aspect of logoic expression, and influence the form side of all manifestation. They focus primarily through *Humanity*.

Esoteric Astrology, pp. 415/6
See also: 'The Science of Triangles'
Ibid, pp. 415/502
and references to The Great Bear, the Pleiades and Sirius,
Ibid, pp. 655/60

2. THE SOLAR SYSTEM

• To the various solar Logoi of the vast constellations that are apparent when we scan the starry heavens, the quality of the Logos of our solar system is seen through the medium of that great thought form He has built by the power of His speech, and which is energised by His particular quality of love. When God speaks, the worlds are made and at this present time He is only in process of speaking. He has not yet concluded what He has to say, and hence the present apparent imperfection. When that great divine phrase or sentence which occupies His thought is brought to a close, we will have a perfect solar system inhabited by perfect existences.

A Treatise on Cosmic Fire, pp. 980/1

• The centre in the cosmic body of the ONE ABOUT WHOM NAUGHT BE SAID of which our solar Logos is the embodied force is the *heart centre.* Here we have one of the clues to the mystery of electricity. The sacred planets, with certain allied etheric spheres within the ring-pass-not, are parts of that heart centre, and are 'petals in the Lotus,' or in the heart centre of that great unknown Existence Who stands to the solar Logos as He in His turn stands to the Heavenly Men Who are His centres, and specially as He stands to the particular Heavenly Man Who is the embodied force of the logoic Heart centre. Therefore, it will be apparent to the careful student that the entire force and energy of the system and its life quality will be that which we call (having perforce to use handicapping, misleading words) LOVE. . . .

This force when rightly directed and properly controlled is the great transmuting agency, which eventually will make of the human unit a Master of the Wisdom, a Lord of Love, a Dragon of Wisdom in lesser degree. *A Treatise on Cosmic Fire,* pp 511/2

• *The Law of Solar Evolution.*

It is, of course, a truism to state that the Law of Solar Evolution is the sum-total of all the lesser activities. We might consider this point in connection with the planetary atom, and the solar atom.

The planetary atom has, as all else in nature, three main activities:

First. It rotates upon its own axis, revolves cyclically within its own ring-pass-not, and thus displays its own inherent energy. What is meant by this phrase? Surely that the milliards of atoms which com-

CELESTIAL DYNAMICS

pose the planetary body (whether dense or subtle) pursue an orbital course around the central energetic positive unit. This dynamic force centre must be considered as subsisting naturally in two locations (if such an unsuitable term is permissible) according to the stage, usage, and particular type of the indwelling planetary entity. . . .One such centre is to be found at the North Pole, and two more are located within the planetary sphere, and frequently the inflow of force or energy to these internal centres (via the polar centre) results in those disasters we call earthquakes, and volcanic eruptions. . . .

The planetary atom revolves upon its axis and comes periodically under influences which produce definite effects. These influences are, among others, those of the moon, and of the two planets which lie nearest to it on either side—nearer and farther away from the Sun. . .

Second, the planetary atom also revolves orbitally around its solar centre. This is its expression of rotary-spiral-cyclic action, and its recognition of the divine central magnet. This brings it under the constant impression of other schemes, each of which produces effects upon the planet. It likewise brings it under the inflowing streams of energy from what are termed the zodiacal constellations which reach the planetary scheme via the great centre, the Sun. . . .

The third activity of the planetary atom is that which carries it through space along with the entire solar system, and which embodies its "drift" or inclination towards the systemic orbit in the heavens.

The solar atom must be considered as pursuing analogous lines of activity and as paralleling on a vast scale the evolution of the planetary atom. The entire solar sphere, the logoic ring-pass-not, rotates upon its axis, and thus all that is included within the sphere is carried in a circular manner through the Heavens. The exact figures of the cycle which covers the vast rotation must remain as yet esoteric, but it may be stated that it approximates one hundred thousand years, being, as might be supposed, controlled by the energy of the first aspect, and therefore of the first Ray. This of itself is sufficient to account for varying and diverse influences which may be traced over vast periods by those with the "seeing eye," for it causes a turning of varying parts of the sphere to the differing zodiacal constellations.

This influence (in connection with the planets) is increased or mitigated according to the place of the planets on their various orbital paths. . . .

Like the planetary atom, the solar atom not only rotates on its axis but likewise spirals in a cyclic fashion through the Heavens. This is a different activity to the drift or progressive dynamic motion through the Heavens. It deals with the revolution of our Sun around a central point and with its relation to the three constellations so oft referred to in this *Treatise:*

The Great Bear.
The Pleiades.
The Sun Sirius.

These three groups of solar bodies are of paramount influence where the spiral cyclic activity of our system is concerned. Just as in the human atom the spiral cyclic activity is egoic and controlled from the egoic body, so in connection with the solar system these three groups are related to the logoic Spiritual Triad, atma-buddhi-manas, and their influence is dominant in connection with solar incarnation, with solar evolution, and with solar progress.

Further, it must be added that the third type of motion to which our system is subjected, that of progress onward, is the result of the united activity of the seven constellations (our solar system forming one of the seven) which form the seven centres of the cosmic Logos. This united activity produces a uniform and steady push (if it might so be expressed) toward a point in the heavens unknown as yet to even the planetary Logoi. *A Treatise on Cosmic Fire,* pp. 1054/9

• The consciousness of the cosmic mental plane is the logoic goal of attainment and that the Sirian Logos is to our solar Logos what the human Ego [or soul] is to the [human] personality.

A Treatise on Cosmic Fire, p. 592

3. THE SUN

• "The sun is neither a Solid nor a Liquid, nor yet a gaseous glow;

but a gigantic ball of electro-magnetic Forces, the store-house of universal Life and Motion, from which the latter pulsate in all directions, feeding the smallest atom as the greatest genius with the same material unto the end of the Maha Yug."

Mahatma Letters to A. P. Sinnett, p, 165.
A Treatise on Cosmic Fire, footnote, p. 311

•*a.* A solar Logos, the Grand Man of the Heavens, is equally spheroidal in shape. His ring-pass-not comprises the entire circumference of the solar system, and all that is included within the sphere of influence of the Sun. The Sun holds a position analogous to the nucleus of life at the centre of the atom. This sphere comprises within its periphery the seven planetary chains with the synthesising three, making the ten of logoic manifestation. The Sun is the physical body of the solar Logos, His body of manifestation, and His life sweeps cycling through the seven schemes in the same sense as the life of a planetary Logos sweeps seven times around His scheme of seven chains. Each chain holds a position analogous to a globe in a planetary chain. Note the beauty of the correspondence, yet withal the lack of detailed analogy.

b. A solar Logos contains within Himself, as the atoms in His body of manifestation, all groups of every kind, from the involutionary group-soul to the egoic groups on the mental plane. He has (for the animating centres of His body) the seven major groups or the seven Heavenly Men, who ray forth Their influence to all parts of the logoic sphere, and who embody within Themselves all lesser lives, the lesser groups, human and deva units, cells, atoms and molecules.

Seen from cosmic levels, the sphere of the Logos can be visualised as a vibrating ball of fire of supernal glory, containing within its circle of influence, the planetary spheres likewise vibrating balls of fire. The Grand Man of the Heavens vibrates to a steadily increasing measure; the entire system is tinctured by a certain color,—the color of the life of the Logos, the One Divine Ray; and the system rotates to a certain measure, which is the key of the great kalpa or solar cycle, and revolves around its central solar pole.

c. The solar Logos is distinguished by His activity on all the planes of the solar system; He is the sumtotal of all manifestation, from the lowest and densest physical atom up to the most radiant and cosmic ethereal Dhyan Chohan. This sevenfold vibratory measure is the key of the lowest cosmic plane, and its rate of rhythm can be felt on the cosmic astral, with a faint response on the cosmic mental. Thus the life of the logoic existence on cosmic levels, may be seen paralleling the life of a man in the three worlds, the lowest of the systemic planes. *A Treatise on Cosmic Fire*, pp. 255/6

• Within the sun, right at its very heart, is a sea of fire or heat, but not a sea of flame. Herein may lie a distinction that perhaps will convey no meaning to some. It is the centre of the sphere, and the point of fiercest internal burning, but has little relation to the flames or burning gases (whatever terms you care to use) that are generally understood to exist whenever the sun is considered. It is the point of fiercest incandescence, and the objective sphere of fire is but the manifestation of that internal combustion. This central heat radiates its warmth to all parts of the system by means of a triple channel, or through its "Rays of Approach" which in their totality express to us the idea of "the heat of the sun."

1. *The akasha,* itself vitalised matter, or substance animated by latent heat.
2. *Electricity*, substance of one polarity, and energised by one of the three aspects logoic. To express it more occultly, substance showing forth the quality of the cosmic Lord Whose energy it is.
3. *Light Rays of pranic aspect*, some of which are being now recognised by the modern scientist. They are but aspects of the latent heat of the sun as it approaches the Earth by a particular line of least resistance.

When the term "channel or ray of approach" is used, it means approach from the centre of solar radiation to the periphery. What is encountered during that approach—such as planetary bodies, for instance—will be affected by the akashic current, the electrical current, or the pranic current in some way, but all of these currents are only the internal fires of the system when viewed from some other point in universal, though not solar, space. It is, therefore, obvious that this matter of fire is as complex as that of the rays. The internal fires of the solar system become external and radiatory when considered from the standpoint of a planet, while the internal fires of the planet will affect a human being as radiation in exactly the same way as the pranic emanations of his etheric body affect another physical body as radiatory. The point to be grasped in all these aspects is that one and all have to do with matter or substance, and not with mind or Spirit. *A Treatise on Cosmic Fire,* pp. 58/60

4. PLANETOIDS–ASTEROIDS

• *The circle of the planetoids.* Students of the Ageless Wisdom are apt to forget that the Life of the Logos manifests itself through those circling spheres which (though not large enough to be regarded as planets) pursue their orbital paths around the solar centre and have their own evolutionary problems and are functioning as part of the solar Body. They are informed—as are the planets—by a cosmic Entity and are under the influence of the Life impulses of the solar Logos as are the greater bodies. The evolutions upon them are analogous to, though not identical with, those of our planet, and they swing through their cycles in the Heavens under the same laws as do the greater planets.

A Treatise on Cosmic Fire, p. 1176

• The true figures in connection with any planetary scheme and its occult activity are not ascertainable by the man who cannot be trusted with the significance of the other planetary bodies (of great number) within the solar ring-pass-not. The entire solar sphere is full of such bodies, characterised by the same features as are the seven and the ten, and each of them in some degree has an effect upon the whole. Figures, therefore, cannot be considered as final until the effect of the lesser planetary bodies upon their immediate neighbours is known, and the extent of their planetary radiation has been gauged. There are more than 115 of such bodies to be reckoned with, and all are at varying stages of vibratory impulse. They have definite orbits, they turn upon their axis, they draw their "life" and substance from the sun, but owing to their relative insignificance, they have not yet been considered factors of moment. This attitude of mind will change when etheric vision is a fact, and the reality of the existence of an etheric double of all that is in manifestation will be recognised by scientists. This fact will be demonstrated towards the close of the century, and, during the early part of the next century a revolution in astronomical circles will occur which will result in the study of the "etheric planets." As these bodies are organs of energy, permeating the dense form, the study of the interaction of solar energy, and the occult "give and take" of planetary bodies will assume a new significance. Certain planetary bodies (both greater and lesser) are "absorbers," others are "radiators," while some are in the stage of demonstrating a dual activity, and are being "transmuted." All these circumstances require to be considered by the initiate who is dealing with cycles.

Figures also must be computed when the effect upon the planets of what are called "asteroids" is known. This is much greater than exoteric science has so far admitted, but the significance of this must eventually be interpreted in terms of energy and on etheric levels.

A Treatise on Cosmic Fire, pp. 793/4

• Unseen Planets: "Not all of the Intra-Mercurial planets, nor yet those in the orbit of Neptune, are yet discovered, although they are strongly suspected. We know that such exist and where they exist; and that there are innumerable planets "burnt out" they say,—in Obscuration we say;—planets in formation and not yet luminous, etc.."...

"When so attached the 'tasimeter'* will afford the possibility not only to measure the heat of the remotest of visible stars, but to detect by their invisible radiations stars that are unseen and otherwise undetectable, hence planets also. The discoverer . . . thinks that if, at any point in a blank space of heavens—a space that appears blank even through a telescope of the highest power—the tasimeter indicates an accession of temperature and does so invariably; this will be a regular proof that the instrument is in range with the stellar body either non-luminous or so distant as to be beyond the reach of telescopic vision. His tasimeter, he says, 'is affected by a wider range of etheric undulations than the Eye can take cognisance of.' Science will Hear sounds from certain planets before she Sees them. This is a Prophecy."—*Mahatma Letters* to A. P. Sinnett, p. 169.

A Treatise on Cosmic Fire, footnote., p. 837
*A tasimeter measures infrared radiation.

5. PLANET EARTH

• Why did the planetary Logos create this world and start the evolutionary, creative process? Only one answer has as yet been permitted to be given. Sanat Kumara has created this planet and all that moves and lives therein in order to bring about a planetary synthesis and an integrated system whereby a tremendous solar revelation can be seen. *The Rays and the Initiations,* p. 717

• ... the seven creative Builders or planetary Logoi of our solar system are embodiments of the will, energy, and magnetic force which streams through Them from the seven solar systems into Their various spheres of activity. Thus, through Their united activity, the organised solar system is produced, whose energies are in constant circulation and whose emerging qualities are balanced and demonstrated throughout the entire system. All parts of the solar system are interdependent; all the forces and energies are in constant flux and mutation; all of them sweep in great pulsations, and through a form of rhythmic breathing, around the entire solar atom; so that the qualities of every solar life, pouring through the seven ray forms, permeate every form within the solar ring-pass-not, and thus link every form with every other form. Note therefore the fact that each of the seven rays or creative Builders embodies the energy, will, love and purpose of the Lord of the solar system, as that Lord in His turn embodies an aspect of the energy, will, love and purpose of the "One About Whom Naught May Be Said." *Esoteric Psychology* I, p. 151

• ... it must be remembered that the sumtotal of human and deva units upon a planet make the *body vital* of a planetary Logos, whilst the sumtotal of lesser lives upon a planet (from the material bodies of men or devas down to the other kingdoms of nature) form His *body corporeal,* and are divisible into two types of such lives:

a. Those on the evolutionary arc, such as in the animal kingdom.
b. Those on the involutionary arc, such as the totality of all elemental material forms within His sphere of influence. All the involutionary lives, as earlier pointed out, form the vehicles for the spirit of the planet, or the planetary entity, who is the sumtotal of the elemental essences in process of involution. He holds a position (in relation to a Heavenly Man) analogous to that held by the different elementals that go to the make-up of man's three bodies, physical, astral and mental, and he is—like all manifesting beings—threefold in his nature, but involutionary. Therefore, man and devas (differentiating the devas from the lesser Builders) form the SOUL of a Heavenly Man. Other lives form his BODY, and it is with body and soul that we are concerned in these two divisions of our thesis on FIRE. One group manifests the fire of matter, the other group the fire of mind, for the devas are the personification of the active universal mind, even though man is considered manasic in a different sense. Man bridges in essence; the devas bridge in matter.

A Treatise on Cosmic Fire, pp. 301/2

• ... one of the mysteries revealed at initiation is that of the logoic centre which our scheme represents, and the type of electrical fire which is flowing through it. The "Seven Brothers," or the seven types of fohatic force, express Themselves through the seven centres, and the One Who is animating our scheme stands revealed at the third

Initiation. It is by knowledge of the nature and quality of the electrical force of our centre, and by realisation of the place our centre holds in the body logoic, that the Hierarchy achieves the aims of evolution. It will be obvious that the Heavenly Man Who stands for the kundalini centre, for instance, will work differently, and have a different purpose and method, from His Brother Who stands for the heart centre in the body logoic, or to the Heavenly Man Who embodies the logoic solar plexus. From this it is apparent that:

a. The type of electrical force
b. The vibratory action
c. The purpose
d. The evolutionary development
e. The dual and triangular interaction

of all the Heavenly Men will differ, and so will the evolutions that form the cells in Their Bodies differ likewise. Little has as yet been revealed anent the types of evolutions which are to be found in the other schemes of our system. Suffice it to say that in all the schemes, on some globe in the scheme, human beings or self-conscious units, are to be found. Conditions of life, environment and form may differ, but the human Hierarchy works in all schemes.

A Treatise on Cosmic Fire, pp. 358/9

• *The Planet.* Deep in the heart of the planet—such a planet as the Earth, for instance—are the internal fires that occupy the central sphere, or the caverns which—filled with incandescent burning—make life upon the globe possible at all. The internal fires of the moon are practically burnt out, and, therefore, she does not shine save through reflection, having no inner fire to blend and merge with light external. These inner fires of the earth can be seen functioning, as in the sun, through three main channels:

1. *Productive substance,* or the matter of the planet vitalised by heat. This heat and matter together act as the mother of all that germinates, and as the protector of all that dwells therein

and thereon. This corresponds to the akasha, the active vitalised matter of the solar system, that nourishes all as does a mother.

2. *Electrical fluid*, a fluid which is latent in the planet though as yet but little recognized. It is perhaps better expressed by the term "animal magnetism." It is the distinctive quality of the atmosphere of a planet, or its electrical ring-pass-not. It is the opposite pole to the solar electrical fluid, and the contact of the two and their correct manipulation is the aim—perhaps unrealised—-of all scientific endeavor at this time.

3. That emanation of the planet which we might term *Planetary Prana*. It is that which is referred to when one speaks of the health-giving qualities of Mother Nature, and which is back of the cry of the modern physician, when he wisely says "Back to the Earth." It is the fluidic emanation of this prana which acts upon the physical body, though in this case not via the etheric body. It is absorbed through the skin purely and the pores are its line of least resistance.

A Treatise on Cosmic Fire, pp. 60/1

• The planetary ray is always the third Ray of Active Intelligence because it conditions our Earth and is of great potency, enabling the human being to "transact his business in the world of planetary physical life." *Esoteric Healing*, pp. 590/1

• As progress is made in the course of taking the higher initiations, it becomes apparent to the initiate that two major streams of energy enter our planetary life:

a. A stream of energy coming from the cosmic mental plane and from that focal point which is to Sanat Kumara what the egoic lotus, the soul, is to the spiritual man; it carries the life principle of our planet and centres itself in Shamballa. From there it is dispersed throughout all forms upon the planet and we call it LIFE. It must be remembered that this life principle embodies or is impregnated with the will and purpose of

THAT which overshadows Sanat Kumara as the soul overshadows the personality.

b. A stream of energy coming from the sun, Sirius; this enters directly into the Hierarchy and carries with it the principle of buddhi, of cosmic love. This, in a mysterious way, is the principle found at the heart of every atom.

The Rays and the Initiations, pp. 414/5

• [Planetary Entity] It might here be noted that the planetary Entity is the sumtotal of all the elemental lives of the lesser Builders functioning as, or forming, the substance of any particular globe in physical objectivity. The mystery of the whole subject lies hidden in three things:

First, the fact that our three planes, physical, astral, and mental, form the dense body of the solar Logos, and are therefore not considered as forming principles.

The second fact is that the lesser "lives" or the elemental essence are the "refuse" of an earlier system, and react to inherent impulses so powerfully that it was only possible to control them through the dynamic will of the Logos, consciously applied. The word "refuse" must be interpreted analogically, and as is understood when it is said that man gathers to himself in each fresh incarnation matter to form his dense physical body which is tinged with the earlier vibrations of preceding incarnations. These "lives" have been gradually drawn in during the entire mahamanvantara as it became safe and possible to control and bend them to the will of the greater Builders. Much of the earlier energy-substance in systemic construction has been passed on into that force-matter which we call that of the lunar Pitris, and its place has gradually been taken by this type of energy, gathered in from the greater sphere in which our Logos has place. The twelve evolutions are after all but the twelve types of energy, manifesting ever in three groups of forces, and again as one group when synthesised during the process of manifestation. They are fourfold in interaction, and have a systemic ebb and flow about which little is known.

Third, the fact of the coming into incarnation of the informing "life" of this low grade substance, who is an entity from a point in the Heavens which may not be mentioned: He embodies influences of a manasic nature, but manas at its very lowest vibration. Perhaps some idea of this may be gathered if it is stated that there is a resemblance between this vibration, or this energising life, and the basic vibration of the solar system preceding this one. We must remember that our basic vibration was the result of the evolutionary process of the entire earlier system. This entity has the same analogous relation to the deva evolution as the mysterious "bridges" which baffle scientists, and which are found between the vegetable and the animal kingdom, and the mineral and the vegetable; they are neither the one nor the other. On a large scale, this "life" or the informing entity of the lower life of the physical plane of the solar system is neither a full exponent of the subconscious life of the earlier system, nor of the elemental life of this; only in the next system will be seen the manifestation of a form of consciousness of a type at present inconceivable to man. Esoterically he is stated to have "neither sight nor hearing"; he is neither deva nor human in essence. He is occultly "blind," utterly unaware; he is capable only of movement, and resembles the foetus in the womb; that which is coming to the birth only the next greater cycle will reveal.

A Treatise on Cosmic Fire, pp. 845/6

• Man, standing as he does at the middle point in evolution, and marking the stage in the evolution of consciousness where a triple awareness is possible,—awareness of individuality, awareness of the forces which are subhuman and which must be controlled, and awareness of a place within the plan and purpose of a greater Man—must, therefore, rightly be regarded as the most important of the evolutions, for through him can be worked out intelligently the laws of group unity for all the three groups, superhuman, human, and subhuman.

Above him stand those who are too pure or, as it is called, "too cold" to be immersed in the matter of the three worlds, below him are found those lives which are too impure (occultly understood) or "too full of burning matter and veiled in smoke" to be able to mount of themselves into regions where stand the unveiled sons of God. Man, therefore, acts as the mediator, and in him and through him can be worked out group methods and laws which—in a later solar system—can form a basic law for unified work. It is this fact which brings about so much of the peculiar trouble and nature of the human kingdom, and it might here be said that on our planet, which is, it must be remembered, one of the "profane" planets, certain experiments in connection with this problem have been undertaken by our planetary Logos. These (if successful) will result in a great expansion of the knowledge of the planetary Logos regarding the laws governing all bodies, and masses. Our planetary Logos has been given the name of the "experimenting divine Physicist." It is this condition which makes the humanity of this planet unique in some respects, for they may be regarded as working out two main problems:

1. The problem of establishing a *conscious* relation, and response, to the animal kingdom.
2. The problem of simultaneously receiving and holding vibrations from superhuman lives and of transmitting them consciously to the subhuman states.

All this has to be accomplished by the units of the human kingdom in full individual consciousness, and the work of each human being might be regarded therefore as having in view the establishment of a sympathetic relation with other human units and with the pitris of the animal kingdom, and also the development of the power to act as the transmitter of energies from greater lives than his own, and to become a transmuting mediating agency.

A Treatise on Cosmic Fire, pp. 1211/3

• In many esoteric books it has been stated and hinted that there has been a mistake, or a serious error, on the part of God Himself, of our planetary Logos, and that this mistake has involved our planet and all that it contains in the visible misery, chaos and suffering. Shall we say that there has been no mistake, but simply a great experiment, of the success or failure of which it is not yet possible to judge? The objective of the experiment might be stated as follows: It is the intent of the planetary Logos to bring about a psychological condition which can best be described as one of "divine lucidity". The work of the psyche, and the goal of the true psychology, is to see life clearly, as it is, and with all that is involved. This does not mean conditions and environment, but Life. This process was begun in the animal kingdom, and will be consummated in the human. These are described in the *Old Commentary* as "the two eyes of Deity, both blind at first, but which later see, though the right eye sees more clearly than the left". The first dim indication of this tendency towards lucidity is seen in the faculty of the plant to turn towards the sun. It is practically non-existent in the mineral kingdom. *Esoteric Psychology* I, pp. 252/3

a. *Appropriation by Planetary Logos of Physical Body*

• . . . in the middle of the Lemurian epoch, approximately eighteen million years ago, occurred a great event which signified, among other things, the following developments:—The Planetary Logos of our earth scheme, one of the Seven Spirits before the throne, took physical incarnation, and, under the form of Sanat Kumara, the Ancient of Days, and the Lord of the World, came down to this dense physical planet and has remained with us ever since. . . .

With the Ancient of Days came a group of other highly evolved Entities, who represent His own individual karmic group and those Beings who are the outcome of the triple nature of the Planetary Logos. If one might so express it They embody the forces emanating from the head, heart, and throat centres, and They came in with Sanat Kumara to form focal points of planetary force for the helping of the great plan for the self-conscious unfoldment of all life. . . .

The result of Their advent, millions of years ago, was stupendous, and its effects are still being felt. Those effects might be enumerated as follows:—The Planetary Logos on His own plane was enabled to take a more direct method in producing the results He desired for working out His plan. As is well known, the planetary scheme, with its dense globe and inner subtler globes, is to the Planetary Logos what the physical body and its subtler bodies are to man.

Hence in illustration it might be said that the coming into incarnation of Sanat Kumara was analogous to the firm grip of self-conscious control that the Ego of a human being takes upon his vehicles when the necessary stage in evolution has been achieved. . . .

The third kingdom of nature, the animal kingdom, had reached a relatively high state of evolution, and animal man was in possession of the earth; he was a being with a powerful physical body, a co-ordinated astral body, or body of sensation and feeling, and a rudimentary germ of mind which might some day form a nucleus of a mental body. . . .

The decision of the Planetary Logos to take a physical vehicle produced an extraordinary stimulation in the evolutionary process, and by His incarnation, and the methods of force distribution He employed, He brought about in a brief cycle of time what would otherwise have been inconceivably slow. The germ of mind in animal man was stimulated. The fourfold lower man,

  a. The physical body in its dual capacity, etheric and dense,
  b. Vitality, life force, or prana,
  c. The astral or emotional body,
  d. The incipient germ of mind,

was co-ordinated and stimulated, and became a fit receptacle for the coming in of the self-conscious entities, those spiritual triads (the reflection of spiritual will, intuition, or wisdom, and higher mind) who had for long ages been waiting for just such a fitting. The fourth, or human kingdom, came thus into being, and the self-conscious, or rational unit, man, began his career.

Another result of the advent of the Hierarchy was a similar, though less recognised development in all the kingdoms of nature. In the mineral kingdom, for instance, certain of the minerals or elements received an added stimulation, and became radioactive, and a mysterious chemical change took place in the vegetable kingdom. This facilitated the bridging process between the vegetable and animal kingdoms, just as the radio-activity of minerals is the method of bridging the gulf between the mineral and vegetable kingdoms. In due course of time scientists will recognise that every kingdom in nature is linked and entered when the units of that kingdom become radioactive. *Initiation, Human and Solar,* pp. 28/32

• Human individualisation, or the emergence of the self-conscious units on the mental plane, is involved in a larger development, for it synchronizes with the appropriation of a dense physical body by the Planetary Logos; this body is composed of matter of our three lower planes. As the etheric centres of the Manasaputras (Planetary Logoi) on the fourth cosmic etheric plane become vitalised, they produce increased activity on the systemic mental plane, the cosmic gaseous, and the consciousness of the Heavenly Man and His life energy begins to make itself felt. Simultaneously, under the Law, mind force or manasic energy pours in from the fifth cosmic plane, the cosmic mental. This dual energy, contacting that which is inherent in the dense physical body of the Logos itself, produces correspondences to the centres upon that plane and the egoic groups appear. They blend in latency the three types of electricity, and are themselves electrical phenomena. They are composed of those atoms, or types of lives, which are a part of the fourth Creative Hierarchy, the aggregate of purely human Monads. Similarly, this triple force, produced by this conscious appropriation by the Heavenly Man, animates deva substance and the dense physical body of the planetary Logos is manifested objectively. *A Treatise on Cosmic Fire,* pp. 690/1

• The pouring in of this force of energy, emanating from the fifth logoic Principle, brings about two things:

> The appropriation by the sevenfold Logos of His dense physical body.
> The appearance on the fifth systemic plane of the causal bodies of the human Monads.
>
> or
>
> For the greater Life it was incarnation.
> For the lesser lives it was individualisation.

CELESTIAL DYNAMICS

This needs pondering upon.

It will, therefore, be apparent to all thinkers why this fifth principle stirred the third aspect into self-conscious activity.

A Treatise on Cosmic Fire, pp. 692/3

6. MOON CHAIN

• It is an interesting occult fact that our Earth should now be in her fifth round, and paralleling the Venusian scheme, but the moon chain of our scheme saw a period of temporary retardation of the evolutionary process of our Heavenly Man; it resulted in a temporary slowing down of His activities, and caused "lost time," if such an expression might reverently be permitted. The Lords of the Dark Face, or the inherent forces of matter for a time achieved success, and only the fifth round of our chain will see their ultimate defeat. The Venusian scheme also had its battleground, but the planetary Logos of that scheme overcame the antagonistic forces, triumphed over material forms, and was consequently in a position—when the right time came—to apply the needed stimulation or an increased fiery vibration to our Earth scheme. The fact that outside aid was called in during the third root-race of this chain, and that the evolution of manas brought about the individualisation, in physical form, of the Avatar, needs to be pondered on. The Divine Manasaputra, the Lord of the World, took form Himself through the driving impulse of manas, inherent in His nature, and in some mysterious way this was aided by another Heavenly Man of another scheme. His co-operation was required.

A Treatise on Cosmic Fire, pp. 392/3

• The Moon chain has in itself a curious occult history, not yet to be disclosed. This differentiates it from the other chains in the scheme and even from any other chain in any scheme. An analogous situation or correspondence will be found in another planetary scheme within the solar system. All this is hidden in the history of one of the solar systems which is united to ours within a cosmic ring-pass-not. Hence the impossibility of yet enlarging upon it. Each Heavenly Man of a scheme is a focal point for the force and power and vibratory life of seven stupendous ENTITIES in exactly the same sense as the seven centres in a human being are the focal points for the influence of a corresponding heavenly Prototype. Our Heavenly Man, therefore, is esoterically allied to one of the seven solar systems, and in this mysterious alliance is hidden the mystery of the moon chain.

Certain brief hints may be given for the due consideration of students:

- The Moon chain was a chain wherein a systemic failure was to be seen.
- It is connected with the lower principles, which H.P.B. has stated are now superseded.
- The sexual misery of this planet finds its origin in the moon failure.
- The progress of evolution on the moon was abruptly disturbed and arrested by the timely interference of the solar Logos. The secret of the suffering in the Earth chain, which makes it merit the name of the Sphere of Suffering, and the mystery of the long and painful watch kept by the SILENT WATCHER, has its origin in the events which brought the moon chain to a terrific culmination. Conditions of agony and of distress such as are found on our planet are found in no such degree in any other scheme.
- The misuse of the vibratory power of a certain centre, and the perversion, or distortion of force to certain erroneous ends, not along the line of evolution, account for much of the moon mystery.
- Certain results, such as the finding of its polar opposite, were hastened unduly on the moon chain, and the consequence was an uneven development and a retardation of the evolution of a certain number of deva and human groups.
- The origin of the feud between the Lords of the Dark Face and the Brotherhood of Light, which found scope for activity in Atlantean days, and during the present root race, can be traced back to the moon chain.

We have here all that it is possible to give out at this time, and much that has hitherto not been permitted publication.

A Treatise on Cosmic Fire, pp. 415/7

• A very pertinent question might here be asked, and though we may not fully explain the mystery, a few suggestive hints may be possible. We might ask: What causes the apparent deadness of the Moon? Is there deva life upon it? Does solar prana have no effect there? What constitutes the difference between the apparently dead Moon, and a live planet, such as the Earth?

Here we touch upon a hidden mystery, of which the solution lies revealed for those who seek, in the fact that human beings and certain groups of devas are no longer found upon the Moon. *Man has not ceased to exist upon the Moon because it is dead and cannot therefore support his life, but the Moon is dead because man and these deva groups have been removed from off its surface and from its sphere of influence.* Man and the devas act on every planet as intermediaries, or as transmitting agencies. Where they are not found, then certain great activities become impossible, and disintegration sets in. The reason for this removal lies in the cosmic Law of Cause and Effect, or cosmic karma, and in the composite, yet individual, history of that one of the Heavenly Men Whose body, the Moon or any other dead planet at any time happened to be. *A Treatise on Cosmic Fire*, pp. 92/3

7. MILKY WAY

• The etheric body is really a net-work of fine channels, which are the component parts of one interlacing fine cord,—one portion of this cord being the magnetic link which unites the physical and the astral bodies and which is snapped or broken after the withdrawal of the etheric body from the dense physical body at the time of death. The silver cord is loosed, as the Bible expresses it and this is the basis of the legend of the fateful sister who cuts the thread of life with the dreaded shears.

The etheric web is composed of the intricate weaving of this vitalised cord, and apart from the seven centres within the web (which correspond to the sacred centres, and of which the spleen is frequently counted as one) it has the two above mentioned, which make—with the spleen—a triangle of activity. The etheric web of the solar system is of an analogous nature, and likewise has its three receptive centres for cosmic prana. The mysterious band in the heavens, which we call the *Milky Way*, is closely connected with cosmic prana, or that cosmic vitality or nourishment which vitalises the solar etheric system.

A Treatise on Cosmic Fire, pp. 98/9

XI. STATES OF MATTER

1. ATOMS–ELECTRONS–IONS

• An atom is a centre of force, a phase of electrical phenomena, a centre of energy, active through its own internal make-up, and giving off energy or heat or radiation.

The Consciousness of the Atom, p. 34

• An Atom -

a. An atom consists of a spheroidal form containing within itself a nucleus of life.

b. An atom contains within itself differentiated molecules, which in their totality form the atom itself. For instance, we are told that the physical atom contains within its periphery fourteen thousand millions of the archetypal atoms, yet these myriads demonstrate as one.

c. An atom is distinguished by activity, and shows forth the qualities of:

a. Rotary motion.
b. Discriminative power.
c. Ability to develop.

d. An atom, we are told, contains within itself three major spirals and seven lesser, which ten are in process of vitalisation, but have not yet attained full activity. Only four are functioning at this stage, and the fifth is in process of development.

e. An atom is governed by the Law of Economy, is coming slowly under the Law of Attraction, and will eventually come under the Law of Synthesis.

f. An atom finds its place within all forms; it is the aggregation of atoms that produces form.

g. Its responsiveness to outer stimulation:

Electrical stimulation, affecting its objective form.
Magnetic stimulation, acting upon its subjective life.
The united effect of the two stimulations, producing consequent internal growth and development.

An atom therefore is distinguished by:
  1. Its spheroidal shape. Its ring-pass-not is definite and seen.
  2. Its internal arrangement, which comprises the sphere of influence of any particular atom.
  3. Its life-activity, or the extent to which the life at the centre animates the atom, a relative thing at this stage.
  4. Its sevenfold inner economy in process of evolution.
  5. Its eventual synthesis internally from the seven into the three.
  6. Its group relation.
  7. Its development of consciousness, or responsiveness.

A Treatise on Cosmic Fire, pp. 246/7

• The subsidiary laws under the Law of Economy are four in number, dealing with the lower quaternary:

  1. *The Law of Vibration,* dealing with the key note or measure of the matter of each plane. By knowledge of this law the material of any plane in its seven divisions can be controlled.
  2. *The Law of Adaptation*, is the law governing the rotary movement of any atom on every plane and subplane.
  3. *The Law of Repulsion*, governs that relationship between atoms, which results in their non-attachment and in their complete freedom from each other; it also keeps them rotating at fixed points from the globe or sphere of opposite polarity.
  4. *The Law of Friction,* governs the heat aspect of any atom, the radiation of an atom, and the effect of that radiation on any other atom.

Every atom of matter can be studied in four aspects, and is governed by one or other, or all of the four above mentioned laws.

  a. An atom vibrates to a certain measure.
  b. It rotates at a certain speed.
  c. It acts and reacts upon its environing atoms.
  d. It adds its quota to the general heat of the atomic system, whatever that may be.

STATES OF MATTER

These general rules relating to atomic bodies can be extended not only to the atoms of the physical plane, but to all spheroidal bodies within the system, and including the system also, regarding it as a cosmic atom.

The tiny atom of the physical plane, a plane itself, a planet, and a solar system all evolve under these rules, and all are governed by the Law of Economy in one of its four aspects.

It might be added in closing, that this law is one that initiates have to master before They can achieve liberation. They have to learn to manipulate matter, and to work with energy or force in matter under this law; they have to utilise matter and energy in order to achieve the liberation of Spirit, and to bring to fruition the purposes of the Logos in the evolutionary process.

A Treatise on Cosmic Fire, pp. 219/20

• If you take these different qualities of the atom—energy, intelligence, ability to select and reject, to attract and repel, sensation, movement, and desire—you have something which is very much like the psychology of a human being, only within a more limited radius and of a more circumscribed degree. Have we not, therefore, really got back to what might be termed the "psyche of the atom"? We have found that the atom is a living entity, a little vibrant world, and that within its sphere of influence other little lives are to be found, and this very much in the same sense that each of us is an entity, or positive nucleus of force or life, holding within our sphere of influence other lesser lives, i.e. the cells of our body. What can be said of us can be said, in degree, of the atom.

The Consciousness of the Atom, pp. 41/2

• Every atom, though termed spheroidal, is more accurately a sphere slightly depressed at one location, that location being the place through which flows the force which animates the matter of the sphere. This is true of all spheres, from the solar down to the atom of matter that we call the cell in the body physical. Through the depression in the physical atom flows the vitalising force from without.

Every atom is both positive and negative; it is receptive or negative where the inflowing force is concerned, and positive or radiatory where its own emanations are concerned, and in connection with its effect upon its environment. *A Treatise on Cosmic Fire*, pp. 155/6

• The negative and positive ions with which the scientist deals are etheric in nature and, therefore, of the physical plane. These unseen particles of substance which can only be traced through their effects and through interference with their activities, are rapidly moving particles in relation to each other and, at the same time, are themselves affected by a greater controlling factor which keeps them so moving.
Esoteric Healing, pp. 369/70

• The activity which holds the electrons gathered around their centre is recognized as identical in nature with that which holds the planets in their orbits around the sun, and between these two divine manifestations the whole range of form is found.
The Light of the Soul, p. 91

• The dense physical forms are an illusion because they are due to the reaction of the eye to those forces about which we have been speaking. Etheric vision, or the power to see energy-substance, is true vision for the human being, just as the etheric is the true form. But until the race is further evolved, the eye is aware of, and responds to the heavier vibration only. Gradually it will shake itself free from the lower and coarser reactions, and become an organ of true vision. It might be of interest here to remember the occult fact that as the atoms in the physical body of the human being pursue their evolution, they pass on and on to ever better forms, and eventually find their place within the eye, first of animals and then of man. This is the highest dense form into which they are built, and marks the consummation of the atom of dense matter. Occultly understood, the eye is formed through the interplay of certain streams of force, of which there are three in the animal, and five in the human being. By their conjunction and interaction, they form what is called "the triple opening" or the "fivefold door" out of which the animal soul or the human spirit can "look out upon the world illusion."

A Treatise on Cosmic Fire, pp. 1096/7

• The soul alone perceives correctly; the soul alone has the power to contact the germ or the principle of Buddhi (in the Christian phraseology, the Christ principle) to be found at the heart of every atom, whether it is the atom of matter as studied in the laboratory of the scientist, whether it is the human atom in the crucible of daily experience, whether it is the planetary atom, within whose ring-pass-not all our kingdoms of nature are found, or the solar atom, God in manifestation through the medium of a solar system.

The Light of the Soul, p. 19

2. SOLID–MATTER AND SUBSTANCE

• Forget not that dense matter is not a principle; it is only that which is responsive to the creative principle.

Externalisation of the Hierarchy, p. 673

• Esoterically speaking, the word "matter" or material is given to all forms in the three worlds; and though the average human being finds it difficult to understand that the medium in which the mental processes take place and that of which all thoughtforms are made is matter from the spiritual angle, yet so it is; substance—technically speaking and esoterically understood—is in reality cosmic etheric matter, or that of which the four higher planes of our seven planes are composed. From the human angle, ability to work with and in the cosmic etheric substance demonstrates first of all when the abstract mind awakens and begins to impress the concrete mind; an intuition is an idea clothed in etheric substance, and the moment a man becomes responsive to those ideas, he can begin to master the techniques of etheric control. All this is, in reality, an aspect of the great creative process: ideas, emanating from the buddhic levels of being (the first or lowest cosmic ether) must be clothed in matter of the abstract levels of the mental plane; then they must be clothed in matter of the concrete mental plane; later, with desire matter, and finally (if they live so long) they assume physical form. An idea which comes from the intuitive levels of the divine consciousness is a true idea. It is noted or apprehended by the man who has, within his equipment, substance of the same quality—for it is the magnetic relation between the man and the idea which has made its apprehension possible. In the great creative process he must give form to the idea, if he possibly can, and thus the creative artist or the creative humanitarian comes into being and the divine creative intention is thereby aided.

Telepathy and the Etheric Vehicle, p. 189

• [Published 1922]: In our talks upon evolution, as I mentioned in the first lecture, we have been dealing somewhat with suppositions, and concerning ourselves with possibilities. Certain things we do know, and certain truths have been ascertained; yet even the conclusions of science, for instance, such as were so much spoken of and insisted upon forty years ago, are no longer regarded as facts, and are not used or promulgated as drastically and as emphatically as they were. Science itself is finding every year that its knowledge is very relative. The more a man grasps and knows, the greater is the horizon which opens up before him. Scientists are now venturing into what are the subtler planes of matter, and therefore into the realms of the unproven, and we should remember that, until recently, science has refused to admit their existence. We are passing beyond the sphere of what has been called "solid matter," into such realms as are inferred when we speak about "centres of energy," "negative and positive force," and "electrical phenomena"; and the emphasis is being laid more and more upon quality rather than upon what has been called substance. The further we look ahead, the wider our speculations become, and the more we attempt to account for telepathic, psychic and other phenomena, the more we shall trespass into the realm of what is now the subjective and the subconscious, and the more we shall be forced to express ourselves in terms of quality or of energy.

The Consciousness of the Atom, pp. 123/4

STATES OF MATTER

• "The Plan is the dynamic substance, providing the content of the reservoir upon which the impressing agent can draw and to which the recipient of the impression must become sensitive."

This sentence requires probably a quite serious readjustment in the thinking of most students. The concept of the Plan as Substance will assuredly be new to them, and new perhaps also to you. It is nevertheless a concept which they must endeavour to grasp. Let me phrase it somewhat differently: The Plan constitutes or is composed of the substance in which the Members of the Hierarchy consistently work. Let us take this important concept and break it up into its component parts for the sake of clarity. I am strongly emphasising these words because this concept is of an importance almost beyond human comprehension, and because its understanding may revise and re-vitalise your entire approach to the Plan, and you will therefore be enabled to work in a fresh and in an entirely new manner:

1. The Plan IS substance. It is essentially substantial energy. And energy is substance and nothing else.

2. The substance (which is the Plan) is dynamic in nature, and is therefore impregnated with the energy of WILL.

3. The Plan constitutes a reservoir of energised substance, held in solution by the WILL of Sanat Kumara and embodying His intangible purpose (intangible to us but not intangible to Him).

4. It is this planetary Substance upon which the "impressing agents" must draw—the Nirmanakayas, the Members of the Hierarchy and the working disciples of the world, plus all spiritual sensitives of a certain degree.

5. Recipients of the desired impression must become sensitive to this substantial energy.

This entire proposition can be referred back to the originating Thinker Who brought our manifested world into being, and Who sequentially and under the Law of Evolution is bringing to fruition the objective of His thinking. In the larger and wider sense, it is that sum total of the ocean of energies in which "we live and move and have our being." This is the sevenfold body of the planetary Logos.

Telepathy and the Etheric Vehicle, pp. 118/20

3. LIQUID

• There is again a very close esoteric connection between the fact lying behind the Biblical words "the Spirit of God moved upon the face of the waters," and the ordered lawful activity of the Great Mother as she performs her work of body-building under the impulse of desire. The true relation between the astral plane and the physical plane will only become apparent as students carefully bear in mind that the astral plane of the solar system is the sixth subplane of the cosmic physical plane, and constitutes the sumtotal of the liquid substance of the logoic physical body. When this is realised, the work of the deva essence takes due place; the factor of desire, or of astral motion, and its reflex action upon the physical body via the sixth subplane will become apparent, and the Great Mother will be seen actively engaged, under the influence of desire, in the work of building, nourishing, and producing that warmth and moisture which make manifestation possible. The Mother is the greatest of the devas, and closely linked with the devas of the waters, for moisture of some kind or another is an essential to all life.

A Treatise on Cosmic Fire, pp. 900/1

• Our planetary Logos is one of the Lords spoken of as being a lesser lord, and more "full of passion" than the higher three. Not even yet is His work completed, and deva substance in its various living orders is not yet brought fully under His control. The deva evolution has far to travel.

If this idea be extended to the solar system, it will be apparent that the astral vehicles of the different planetary Logoi differ. This difference is necessarily dependent upon *Their cosmic astral life which directly affects the systemic astral, or the physical liquid subplane of the cosmic physical.* This is a point but little realised. The dense physical body of the planetary Logos exists, as we know, in a threefold condition—dense, liquid, and gaseous—and each is acted upon directly from the corresponding cosmic plane. The condition of

STATES OF MATTER

the various physical planets will some day be found to be dependent upon this fact.

When the psychic nature of the planetary Logos is understood (which knowledge is entered upon after initiation, being a part of the Wisdom) the nature of the different schemes, as regards their watery aspect, for instance, will be found to be connected with a particular astral state. As the initiate progresses in wisdom, he intuitively comprehends the essential nature of the seven groups, or of the logoic Septenate, which is that concerning their colour or quality. This colour or quality is dependent upon the psychic nature of any particular planetary Logos, and His emotional or desire nature can thereby be somewhat studied by the initiate. This will lead eventually to a scientific consideration of the effect of this nature upon His dense physical body, and particularly that portion of it which we call the astral plane, the liquid sub-plane of the cosmic physical plane. A reflection of this (or a further working out, if that term is preferred) is found in the liquid portions of the physical planet.

A Treatise on Cosmic Fire, pp. 673/4

4. GASEOUS

• Speaking generally, it must be remembered that the earth devas of densest matter become, in the course of evolution, the devas of the waters, and find their way eventually on to the astral plane, the cosmic liquid; the devas of the waters of the physical plane find their way, through service, on to the gaseous subplane, and then to the cosmic gaseous, becoming the devas of the mental plane. This literally and occultly constitutes the transmutation of desire into thought.

A Treatise on Cosmic Fire, p. 903

• The devas of the lower mental levels in relation to man work through the mental unit, and are, roughly speaking, divided into four groups, being in fact the first condensation of the threefold lower body of man. They form part of his lunar body. They are directly linked with the highest spiritual essences, and represent the lowest manifestation of force emanating from the cosmic mental plane, and finding its link with the human Hierarchy through the mental units. They are the gaseous devas of the logoic physical body.

A Treatise on Cosmic Fire, pp. 681/2

XII. ATOMIC ENERGY

• THE RELEASE OF ATOMIC ENERGY

August 9, 1945

I would like at this time to touch upon the greatest spiritual event which has taken place since the fourth kingdom of nature, the human kingdom, appeared. I refer to the release of atomic energy, as related in the newspapers this week, August 6, 1945, in connection with the bombing of Japan.

Some years ago I told you that the new era would be ushered in by the scientists of the world and that the inauguration of the kingdom of God on Earth would be heralded by means of successful scientific investigation. By this first step in the releasing of the energy of the atom this has been accomplished, and my prophecy has been justified during this momentous year of our Lord 1945. . . . Certain ideas and suggested thoughts may be of real value here and enable you to see this stupendous event in better perspective.

. . . all great discoveries, such as those made in connection with astronomy or in relation to the laws of nature or involving such a revelation as that of radio-activity or the epoch-making event announced this week concerning the first steps taken in the harnessing of cosmic energy, are ever the result of inner pressure emanating from Forces and Lives found in high Places. Such inner pressures themselves function under the laws of the Spirit and not just under what you call natural laws; they are the result of the impelling work of certain great Lives, working in connection with the third aspect of divinity, that of active intelligence, and are concerned with the substance or matter aspect of manifestation. Such activities are motivated from Shamballa. This activity is set in motion by these Lives, working on Their high plane, and it gradually causes a reaction in the various departments of the Hierarchy, particularly those working under third, fifth and seventh ray Masters. Eventually, disciples upon the physical levels of activity become aware of the inner ferment, and this happens either consciously or unconsciously. They become "impressed," and the scientific work is then started and carried through into the stages of experimentation and final success.

The Externalisation of the Hierarchy, pp. 491/2

• The war was won by the Forces of Light when the mental potency of the forces of evil was overcome and the "energy of the future" was directed or impelled by Those Who were seeking the higher human values and the spiritual good of mankind. Four factors lie behind the momentous happening of the release of this form of atomic energy, through the medium of what is erroneously and unscientifically called the "splitting of the atom." There are other factors, but you may find the following four of real interest:

a. There was a clearly directed inflow of extra-planetary energy released by the Lords of Liberation, to Whom invocation had been successfully made; through the impact of this energy upon the atomic substance being dealt with by the investigating scientists, changes were brought about which enabled them to achieve success. The experiments being carried forward were therefore both subjective and objective.

b. A concerted effort was made by a number of disciples who were working in fifth and seventh ray ashrams, and this enabled them to impress lesser disciples in the scientific field and helped them to surmount the well-nigh insuperable difficulties with which they were confronted.

c. There was also a weakening of the tension which had hitherto successfully held the forces of evil together, and a growing inability of the evil group at the head of the Axis Powers to surmount the incidental war fatigue. This brought about, first of all, a steady deterioration of their minds, and then of their brains and nervous systems. . . .

d. Another factor was the constant, invocative demand and the prayers (articulate and inarticulate) of humanity itself. Men, impelled largely by fear and the innate mobilising of the human spirit against slavery, reached such a pitch of demanding energy that a channel was created which greatly facilitated the work of the Hierarchy, under the direct influence of the Lords of Liberation.

The release of the energy of the atom is as yet in an extremely embryonic stage; humanity little knows the extent or the nature of the energies which have been tapped and released. There are many types of atoms, constituting the "world substance"; each can release its own type of force; this is one of the secrets which the new age will in time reveal, but a good and sound beginning has been made. I would call your attention to the words, "the liberation of energy." It is *liberation* which is the keynote of the new era, just as it has ever been the keynote of the spiritually oriented aspirant. This liberation has started by the release of an aspect of matter and the freeing of some of the soul forces within the atom. This has been, for matter itself, a great and potent initiation, paralleling those initiations which liberate or release the souls of men.

In this process of planetary initiation humanity has carried its work as the world saviour down into the world of substance, and has affected those primary units of life of which all forms are made.

You will now understand the meaning of the words used by so many of you in the second of the Great Invocations: *The hour of service of the saving force has now arrived.* This "saving force" is the energy which science has released into the world for the destruction, first of all, of those who continue (if they do) to defy the Forces of Light working through the United Nations. Then—as time goes on—this liberated energy will usher in the new civilisation, the new and better world and the finer, more spiritual conditions. The highest dreams of those who love their fellowmen can become practical possibilities through the right use of this liberated energy, if the real values are taught, emphasised and applied to daily living. This "saving force" has now been made available by science, and my earlier prophecy substantiated. [See 'Predictions' below.]

As I said above, the first use of this energy has been material destruction; this was inevitable and desirable; old forms (obstructing the good) have had to be destroyed; the wrecking and disappearance of that which is bad and undesirable must ever precede the building of the good and desirable and the longed-for emergence of that which is new and better.

The constructive use of this energy and its harnessing for the betterment of humanity is its real purpose; this living energy of substance itself, hitherto shut up within the atom and imprisoned in these ultimate forms of life, can be turned wholly into that which is good and can bring about such a revolutionising of the modes of human experience that (from one angle alone) it will necessitate and bring about an entirely new economic world structure.

It lies in the hands of the United Nations to protect this released energy from misuse and to see that its power is not prostituted to selfish ends and purely material purposes. It is a "saving force" and has in it the potency of rebuilding, of rehabilitation and of reconstruction. Its right use can abolish destitution, bring civilised comfort (and not useless luxury) to all upon our planet; its expression in forms of right living, if motivated by right human relations, will produce beauty, warmth, colour, the abolition of the present forms of disease, the withdrawal of mankind from all activities which involve living or working underground, and will bring to an end all human slavery, all need to work or fight for possessions and things, and will render possible a state of life which will leave man free to pursue the higher aims of the Spirit. The prostituting of life to the task of providing the bare necessities or to making it possible for a few rich and privileged people to have too much when others have too little, will come to an end; men everywhere can now be released into a state of life which will give them leisure and time to follow spiritual objectives, to realise richer cultural life, and to attain a broader mental perspective. . . .

These few suggestions will give you much food for thought and real ground for happy, confident, forward thinking. Organise now for the goodwill work. The future of the world lies in the hands of the men of goodwill and in those who have unselfish purpose everywhere. This release of energy will eventually make money, as we know it, of

ATOMIC ENERGY

no moment whatsoever; money has proved itself (owing to man's limitations) a producer of evil and the sower of dissension and discontent in the world. This new released energy can prove itself a "saving force" for all mankind, releasing from poverty, ugliness, degradation, slavery and despair; it will destroy the great monopolies, take the curse out of labour, and open the door into that golden age for which all men wait. It will level all the artificial layers of modern society and liberate men from the constant anxiety and gruelling toil which have been responsible for so much disease and death. When these new and better conditions are established, then men will be free to live and move in beauty and to seek the "Lighted Way."

The Externalisation of the Hierarchy, pp. 495/500

1. PREDICTIONS

• [Published in 1934]: It might be noted here that three great discoveries are imminent and during the next two generations will revolutionize modern thought and life.

One is already sensed and is the subject of experiment and investigation, the releasing of the energy of the atom. This will completely change the economic and political situation in the world, for the latter is largely dependent upon the former. Our mechanical civilization will be simplified, and an era ushered in which will be free from the incubus of money (its possession and its non-possession), and the human family will recognize universally its status as a bridging kingdom between the three lower kingdoms of nature and the fifth or spiritual kingdom. There will be time and freedom for a soul culture which will supersede our modern methods of education, and the significance of soul powers and the development of the superhuman consciousness will engross the attention of educators and students everywhere.

A second discovery will grow out of the present investigations as to light and color. The effect of color on people, animals and units in the vegetable kingdom will be studied and the result of those studies will be the development of etheric vision or the power to see the next grade of matter with the strictly physical eye. Increasingly will people think and talk in terms of light, and the effect of the coming developments in this department of human thought will be triple.

a. People will possess etheric vision.

b. The vital or etheric body, lying as the inner structure of the outer forms, will be seen and noted and studied in all kingdoms of nature.

c. This will break down all barriers of race and all distinctions of color; the essential brotherhood of man will be established. We shall see each other and all forms of divine manifestation as light units of varying degrees of brightness and shall talk and think increasingly in terms of electricity, of voltage, of intensity and of power. The age and status of men, in regard to the ladder of evolution, will be noted and become objectively apparent, the relative capacities of old souls, and young souls will be recognized, thereby re-establishing on earth the rule of the enlightened.

Note here, that these developments will be the work of the scientists of the next two generations and the result of their efforts. Their work with the atom of substance, and their investigations in the realm of electricity, of light and of power, must inevitably demonstrate the relation between forms, which is another term for brotherhood, and the fact of the soul, the inner light and radiance of all forms.

The third development, which will be the last probably to take place, will be more strictly in the realm of what the occultists call magic. It will grow out of the study of sound and the effect of sound and will put into man's hands a tremendous instrument in the world of creation. Through the use of sound the scientist of the future will bring about his results; through sound, a new field of discovery will open up; the sound which every form in all kingdoms of nature gives forth will be studied and known and changes will be brought about and new forms developed through its medium. One hint only may I give here and that is, that the release of energy in the atom is linked to this new coming science of sound.

The significance of what has happened in the world during the last century in the realm of sound is not appreciated yet nor understood. Terrific effects are however being produced by the unbelievably increased noise and sound emanating from the planet at this time. The roar of machines, the rumble of the transportation mechanisms in all parts of the world—trains, vessels and airplanes—the focussing of the sounds of men in such congested areas as the great cities, and, at this time, the universal use of the radio bringing musical sounds into every home and into street life are producing effects upon the bodies of men and upon all forms of life everywhere which will become apparent only as time elapses. Some forms of life in the animal kingdom, but primarily in the vegetable kingdom, will disappear and the response of the human mechanism to this world of sound, uproar and music in which it will increasingly find itself will be most interesting.

These three developments will usher in the new age, will produce in this transition period the needed changes, and will inaugurate a new era wherein brotherhood will be the keynote, for it will be a demonstrated fact in nature. It will be an age wherein men will walk in the light, for it will be a world of recognized inner radiance, wherein the work of the world will be carried forward through the medium of sound, and eventually through the use of words of power and the work of the trained magician. These trained workers in substance, understanding the nature of matter, seeing always in terms of light and comprehending the purpose of sound will bring about those structural changes and those material transformations which will establish a civilization adequate for the work of the coming race. This work will be that of the conscious unification of the soul and its vehicle of manifestation. Those cultural methods also which will take the undeveloped of the race and carry them forward to a better manifestation, and a truer expression of themselves will be established and this it is the privilege of the coming generation of scientific investigators to bring about.

The outstanding characteristic, however, of the coming cycle will be an outgrowth of psychology. It will be the emergence of a new factor from the standpoint of the modern psychologist of the materialistic school and will involve the recognition of the soul.

A Treatise on White Magic, pp. 333/6

XIII. PHYSICAL FORCES OF EVOLUTION

1. SOUND–LIGHT–VIBRATION–FORM

- RULE IV. *Sound, light, vibration, and the form blend and merge, and thus the work is one. It proceedeth under the law, and naught can hinder now the work from going forward. The man breathes deeply. He concentrates his forces, and drives the thought-form from him.*   A Treatise on Cosmic Fire, p. 1002

• We have touched upon two words of significance in the fourth Rule,—sound and light,—and one paramount idea emerges. The soul is to be known as light, as the revealer, whilst the Spirit aspect will later be recognized as sound. Complete light and illumination is the right of the disciple who attains to the third initiation, whilst the true comprehension of the sound, of the triple AUM, the synthesizing factor in manifestation appears only to the one who stands master of the three worlds.

The word *vibration* must next engage our attention but it may not be dissociated from the next word in the sequence *form*. Vibration, the effect of divine activity, is two-fold. There is the first effect in which the vibration (issuing from the realm of subjectivity in response to sound and light) produces response in matter, and therefore attracts or calls together the atoms out of which molecules, cells, organisms and finally the integrated form can be built. This effected, the aspect of vibration is to be noted as a duality.

The form, through the medium of the five senses, becomes aware of the vibratory aspect of all forms in the environment wherein it, itself, is a functioning entity. Later, in time and space, that functioning form becomes increasingly aware of its own interior vibration, and by tracing back that vibration to its originating source becomes aware of the Self, and later of the Kingdom of the Self. Humanity as a whole is aware of its environment and, through the information conveyed by the sense of sight, hearing, touch, taste and smell, the phenomenal world, the outer garment of God, is known, and communication between the Self and what we call the natural world is set up.

A Treatise on White Magic, pp. 145/6

• The second word of importance in this fourth Rule is the word *light*. First the sound and then the first effect of sound, the pouring forth of light, causing the revelation of the thought form.

Light is known by what is revealed. The absence of light produces the fading away, into apparent non-existence, of the phenomenal world.

The thought form created by the Sound is intended to be a source of revelation. It must reveal truth, and bring an aspect of reality to the cognisance of the onlooker. Hence the second quality of the thought form in its highest use is that it brings light to those who need it, to those who walk in darkness.

I deal not here with light as the soul, cosmically or individually. I touch not upon light as the universal second aspect of divinity. I seek only in these Instructions to deal with that aspect of truth which will make the aspirant a practical worker, and so enable him to work with intelligence. His main work (and increasingly he will find this to be so) is to create thought-forms to carry revelation to thinking human beings. To do this he must work occultly, and through the sound of his breathed forth work, through the truth revealed in form, will he carry light and illumination into the dark places of the earth.

Then he finally makes his thought form live through the power of his own assurance, spiritual understanding and vitality. Thus the significance of the third word, *vibration*, appears. His message is heard, for it is sounded forth; it carries illumination, for it conveys the Truth and reveals Reality; it is of vital import, for it vibrates with the life of its creator, and is held in being as long as his thought and sound and intelligence animate it. This is true of a message, of an organization, and of all forms of life, which are but the embodied ideas of a cosmic or a human creator.

Students would find it of value to take these three vital words and trace their relation to all embodied thought forms—a cosmos, a plane, a kingdom in nature, a race, a nation, a human being. Consider the diverse groups of creating agencies—solar Logoi, solar Angels, human beings, and others. Consider the spheres of the creative process and see how true the *Old Commentary* is when it says:

> "The sound reverberated amidst varying wheels of uncreated matter; and lo, the sun and all the lesser wheels appeared. The light shone forth amidst the many wheels, and thus the many forms of God, the diverse aspects of his radiant robe blazed forth.
>
> "The vibrant palpitating wheels turned over. Life, in its many stages and in its many grades commenced the process of unfolding, and lo, the law began to work. Forms arose, and disappeared, but life moved on. Kingdoms arose, holding their many forms which drew together, turned together, and later separated, but still the life moved on.
>
> "Mankind, hiding the Son of God, the Word incarnate, broke forth into the light of revelation. Races appeared and disappeared. The forms, veiling the radiant soul, emerged, achieved their purpose and vanished into night, but lo, the life moved on, blended this time with light. Life merged with light, both blending to reveal a beauty and a power, an active liberating force, a wisdom and a love that we call a Son of God.
>
> "Through the many Sons of God, who in their inmost centre are but one, God in his Fatherhood is known. Yet still that lighted life moved on to a dread point of power, of force creative, concerning which we say: It is the All, the Container of the Universe, the persistent centre of the Spheres, the One."

A Treatise on White Magic, pp. 144/5

2. SOUND

• Sound, the basis of existence; sound, the method of being; sound, the final unifier; sound therefore realised as the *raison d'être*, as the method of evolution, and therefore as beatitude.

A Treatise on Cosmic Fire, p. 192

• First, let us repeat the truism that the worlds are the effect of sound. First life, then matter; later the attraction of the matter to the life for purposes of its manifestation and expression, and the orderly arrangement of that matter into the necessary forms. Sound formed the allying factor, the propelling impulse, and the attractive medium. Sound, in an occult and deeply metaphysical sense, stands for that which we term "the relation between", and is the creating intermediary, the linking third factor in the process of manifestation. It is the akasha. On the higher planes it is the agent of the great Entity Who wields the cosmic law of gravitation in its relation to our solar system, whilst on the lower planes it demonstrates as the astral light, the great agent of reflection, that fixes and perpetuates on its vibratory bosom the past, the present and the future, or that which we call Time. In direct relation to the lower vehicle it manifests as electricity, prana and the magnetic fluid. A simplification of the idea may come to you perhaps in the recognition of sound as the agent of the law of attraction and repulsion.   *Letters on Occult Meditation,* pp. 52/3

• *The Solar Words* - The basis of all manifested phenomena is the enunciated sound, or the Word spoken with power, that is, with the full purpose of the will behind it. Herein, as is known, lies the value of meditation, for meditation produces eventually that inner dynamic purpose and recollection, or that internal ideation which must invariably precede the uttering of any creative sound.

When it is said that the Logos produced the worlds through meditation it means that within His own centre of consciousness there was a period wherein He brooded over and meditated upon the purposes and plans He had in view; wherein He visualised to Himself the entire world process as a perfected whole, seeing the end from the beginning and being aware of the detail of the consummated sphere. Then, when His meditation was concluded, and the whole completed as a picture before His inner vision, He brought into use a certain Word of Power which had been committed to Him by the *One about Whom naught may be said*, the Logos of the cosmic scheme of which our system is but a part. With cosmic and Logoic initiations we are not concerned, except in so far as the human initiations reflect their stupendous prototypes, but it is of interest to the student to realise that just as at each initiation some Word of Power is committed to the initiate, so similarly to the Logos was committed the great Word of Power which produced our solar system, that Word which is called the "Sacred Word," or AUM. It must be here remembered that this sound AUM is man's endeavour to reproduce on an infinitesimally small scale the cosmic triple sound whereby creation was made possible. The Words of Power of all degrees have a triple sequence.

First. They are sounded by some fully self-conscious entity, and this invariably takes place after a period of deliberation or meditation wherein the purpose in toto is visualised.

Second. They affect the deva kingdom and produce the creation of forms. This effect is dual in character—

a. The devas on the evolutionary path, the great builders of the solar system, and those under them who have passed the human stage respond to the sound of the Word, and with conscious realisation collaborate with the one who has breathed it forth, and thus the work is carried out.
b. The devas on the involutionary arc, the lesser builders, who have not passed through the human stage, also respond to the sound, but unconsciously, or perforce, and through the power of the initiated vibrations build the required forms out of their own substance.

Third. They act as a stabilising factor, and as long as the force of the sound persists, the forms cohere. When the Logos, for instance, finishes the sounding of the sacred AUM, and the vibration ceases, then disintegration of the forms will ensue. So with the Planetary Logos, and thus on down the scale.

Initiation, Human and Solar, pp. 150/1

• [The Ego builds the human body]: The first step of the Ego towards producing a "shadow" is expressed in the words "The Ego sounds his note." He utters his voice, and (as in the logoic process) the lesser "army of the voice" responds immediately to it. According to the tone and quality of the voice, so is the nature of the responsive agents. According to the depth or height of the note, and according to its volume, so is the status or grade of building deva which replies to the call. This egoic note produces, therefore, certain effects:

It sweeps into activity devas who proceed to transmit the sound. They utter a word.

It reaches the listening devas of the second grade who take up the word and proceed to elaborate it into what might be called a mantric phrase. The building process definitely begins in a sequential threefold manner. The mental body begins to co-ordinate in three stages. All the building stages overlap. When, for instance, the co-ordination of the mental body is in its second stage, the first stage of astral concretion begins. This is carried on for seven stages (three major and four minor) which overlap in an intricate fashion. Again, when the second stage is reached, a vibration is produced which awakens response in etheric matter on the physical plane, and the builders of the etheric double commence their activity. Again the process is repeated. When the second stage of the work of these etheric devas is begun, *conception takes place upon the physical plane.* This is a very important point to be remembered, for it brings the entire process of human birth definitely into line with established karmic law. It shows the close connection between that which is subjective and that which is tangible and seen. The building of the physical body proceeds like that of the three stages during the prenatal period:

a. The work of the building devas during the three and a half months prior to the realisation of life. This period sees the third stage of the building of the etheric body entered upon.
b. The building work of the next three and a half months of the gestation period.
c. The final process of concretion carried on through the remaining two months.

Students will here find it interesting to trace out the correspondence in this method of producing evolutionary manifestation in a planetary scheme with its rounds and races, and in a solar system with its manvantaras and greater cycles.

In summing up this very cursory outline, the work of the etheric devas does not cease at the birth of the man, but is continued likewise in three stages, which find a close analogy in the life period of a solar system.

First, their work is directed to the steady increase of the human physical vehicle, so that it may follow accurately the lines of growth of the two subtler bodies. This is carried on till maturity is reached. The next stage is that in which their work consists largely of repair work, and the preservation of the body during the years of full manhood so that it can measure up to the purpose of the subjective life. This purpose necessarily varies according to the stage of development of the man. Finally comes the stage when the work of building ceases. The vitality in the etheric body waxes dim, and the processes of destruction begin. The Ego begins to call in his forces. The "sound" becomes faint and dim; there is less and less volume for the transmitters to pass on, and the initial vibration gets fainter and fainter. The period of obscuration comes in. First the physical body waxes weak and useless; then the Ego withdraws from the centres, and functions for a few hours in the etheric double. This in turn is devitalised, and so the process is carried on till one by one the sheaths are discarded and the egoic "shadow" is dissipated.

A Treatise on Cosmic Fire, pp. 937/9

• A Master of the Wisdom is He Who has entrusted to Him, by virtue of work accomplished, certain Words of Power. By means of these Words He wields the law over other evolutions than the human, and through them He co-operates with the activity aspect of the Logos. Thus He blends His consciousness with that of the third Logos. Through these Words He assists with the building work, and the cohesive manipulating endeavor of the second Logos, and comprehends the inner working of the law of gravitation (for attraction and repulsion) that governs all the functions of the second aspect logoic. Through these Words He co-operates with the work of the first Logos, and learns, as He takes the sixth and seventh Initiations (which is not always done) the meaning of Will as applied in the system. These Words are imparted orally, and through clairvoyant faculty but must be found by the Initiate Himself, by the use of atma and as He attains atmic consciousness.... When atmic consciousness is developing by means of the intuition, the Initiate can contact the stores of knowledge inherent in the Monad, and thus learn the Words of Power.

Letters on Occult Meditation, pp. 263/4

• By the means of one-pointed meditation upon the relationship between the akasha and sound, an organ for spiritual hearing will be developed.

To understand this sutra, it is essential that certain relationships are comprehended—relationships between matter, the senses and the one who experiences.

The Christian believes that "all things were made by the word of God." The oriental believer holds that sound was the originating factor in the creative process and both teach that this word or sound is descriptive of the second Person of the divine Trinity.

This sound or word threw into peculiar activity the matter of the solar system, and was preceded by the breath of the Father which started the original motion or vibration.

First, therefore, the breath (pneuma or spirit) impinging upon primordial substance and setting up a pulsation, a vibration, a rhythm. Then the word or sound, causing the pulsating vibrating substance to take form or shape, and thus bringing about the incarnation of the second Person of the cosmic Trinity, the Son of God, the Macrocosm.

This process eventuated in the seven planes of manifestation, the spheres wherein seven states of consciousness are possible. All of these are characterized by certain qualities and differentiated from each other by specific vibrating capacities and called by certain terms.

The Light of the Soul, pp. 333/4

• The vibration initiated by the Sound, which is the expression of the Law of Synthesis, is succeeded by the Voice or Word, and that

Word as it progresses outward from the centre to the periphery (for, occultly understood, the Word is "spoken from the Heart") becomes

a. A phrase.

b. Phrases.

c. Sentences.

d. Speech.

e. The myriad sounds of nature.

Each of these terms can be explained in terms of attractive energy, and this attractive energy is likewise the demonstration of the life of an Existence of some grade or other.

"God speaks and the forms are made."

A Treatise on Cosmic Fire, p. 1188

• Of the various karmic agencies wielded by man in the way of molding himself and surroundings, sound or speech is the most important, for, to speak is to work in ether which of course rules the lower quaternary of elements, air, fire, water and earth. Human sound or language contains therefore all the elements required to move the different classes of Devas and those elements are of course the vowels and the consonants. The details of the philosophy of sound in its relation to the devas who preside over the subtle world, belong to the domain of true Mantra Sastra which of course is in the hands of the knowers." *Some Thoughts on the Gita,* p. 72.

A Treatise on Cosmic Fire, footnote, p. 193

• It is a platitude to say that the true meaning of "psychology" is the "word of the soul." It is the sound, producing an effect in matter, which a particular ray may make. This is in some ways a difficult way of expressing it, but if it is realised that each of the seven rays emits its own sound, and in so doing sets in motion those forces which must work in unison with it, the entire question of man's free will, of his eternal destiny and of his power to be self-assertive comes up for solution. *Esoteric Psychology* I, p. 8

See also: 'The Creative Use of Sound',
A Treatise on White Magic, pp. 125/49

PHYSICAL FORCES OF EVOLUTION

a. *Communication*

• The whole planetary system is in reality a vast interlocking, inter-dependent and inter-related complexity of vehicles communicating or responsive to communication.

Telepathy and the Etheric Vehicle, p. 83

• The mental plane is, as H. P. B. has pointed out, the vastest of all the planes with which we are concerned. It is the key plane of the solar system. It is the pivotal plane upon which the great Wheel turns. It is the meeting place of the three lines of evolution and has been for this reason esoterically termed "*the council chamber of the Three Divinities.*" On this plane, the three Persons of the logoic Trinity meet in united work. Below two Persons may be seen associated; above another duality functions, but only on this plane do the Three make an at-one-ment.

All the Logoi of the differing schemes are expressing Themselves upon this plane. There are certain schemes in the system which find their lowest manifestation on this plane, and have no physical body such as the Earth, and the other dense planets. They exist through the medium of gaseous matter, and their spheres of manifestation are simply composed of the four cosmic ethers and the cosmic gaseous. But all the great Lives of the solar system do possess bodies of our systemic mental matter, and therefore on that plane communication be*tween all these Entities becomes a possibility. This fact is the basis of occult realisation, and the true ground for the at-one-ment.* Matter of the abstract levels of the mental plane enters into the composition of the vehicles for all these greater Existences and through the medium of this energised substance each can get *en rapport* with each, no matter what Their individual goal of attainment may be. The units, therefore, in Their bodies can equally get in touch with all other Egos and groups once they have achieved the consciousness of the mental plane (causal consciousness) and know the varying group "keys," the group tones and colors. . . .

Therefore, if one might venture to express an abstraction and a state of consciousness in terms of time and space, and through the limitation of language, it might be stated that on egoic levels, or on the three higher subplanes of the mental plane, there exists a channel of communication, based on similarity of vibration and oneness of endeavour, between every one of the planetary schemes, within the solar ring-pass-not. Here, and here alone (as regards the three worlds and the human kingdom), becomes possible the establishing of egoic relationships and the transmission of thought substance between

a. Units and egoic groups.
b. Groups and other groups.
c. Greater groups with still greater or with lesser ones.
d. Egos in one planetary scheme with those in another.

The Ah-hi, the greater Builders, Who are the Lords working out the will of the solar Logos, mainly use two planes for communication with each other and with Their cohorts:

First, *the second plane,* where They communicate by means of a spiritual medium incomprehensible to man at present.
Second, *the mental plane,* where They communicate with all lesser lives by means of a type of mental telepathy.
A Treatise on Cosmic Fire, pp. 848/9, 50/1

b. *Music of the Spheres*

• The Sound which was the first indication of the activity of the planetary Logos is not a word, but a full reverberating sound, holding within itself all other sounds, all chords and certain musical tones (which have been given the name of the "music of the spheres") and dissonances, unknown as yet to the modern ear. It is this Sound which the "Rising One" must learn to recognise, and to which he must respond not only by means of the sense of hearing and its higher correspondences, but through a response from every part and aspect of the form nature in the three worlds. *Esoteric Healing,* pp. 688/9

• The seven planes of Divine Manifestation, or the seven major planes of our system, are but the seven subplanes of the lowest cosmic plane. The seven Rays of which we hear so much, and which hold so much of interest and of mystery, are likewise but the seven sub-rays of one cosmic Ray. The twelve creative Hierarchies are themselves but subsidiary branches of one cosmic Hierarchy. They form but one chord in the cosmic symphony. When that sevenfold cosmic chord, of which we form so humble a part, reverberates in synthetic perfection, then, and only then, will come comprehension of the words in the Book of Job: "The morning stars sang together." Dissonance yet sounds forth, and discord arises from many systems, but in the progression of the aeons an ordered harmony will eventuate, and the day will dawn when (if we dare speak of eternities in the terms of time) the sound of the perfected universe will resound to the uttermost bounds of the furthest constellation. Then will be known the mystery of "the marriage song of the heavens."

Initiation, Human and Solar, pp. 3/4

• Forget not that sound permeates all forms; the planet itself has its own note or sound; each minute atom also has its sound; each form can be evoked into music and each human being has his peculiar chord and all chords contribute to the great symphony which the Hierarchy and Humanity are playing, and playing now. Every spiritual group has its own tune (if I may employ so inappropriate a word) and the groups which are in process of collaborating with the Hierarchy make music ceaselessly. This rhythm of sound and this myriad of chords and notes blend with the music of the Hierarchy itself and this is a steadily enriching symphony; as the centuries slip away, all these sounds slowly unite and are resolved into each other until some day the planetary symphony which Sanat Kumara is composing will be completed and our Earth will then make a notable contribution to the great chords of the solar system—and this is a part, intrinsic and real, of the music of the spheres. Then, as the Bible says, the Sons of God, the planetary Logoi, will sing together. This, my brother, will be the result of right breathing, of controlled and organised rhythm, of true pure thought and of the correct relation between all parts of the chorus.

Glamour: A World Problem, pp. 259/60

• Then will our spoken word create a thought form which will embody the idea we have in our minds. Then too will our words carry no discord, but will add their quota to that great harmonizing chord or unifying word which it is the function of mankind ultimately to utter. Wrong speech separates, and it is interesting to bear in mind that the word, the symbol of unity, is divine, whereas speech in its many diversifications is human. *A Treatise on White Magic*, p. 143

3. LIGHT

• It must ever be borne in mind that the great theme of LIGHT underlies our entire planetary purpose. The full expression of perfect LIGHT, occultly understood, is the engrossing life-purpose of our planetary Logos. Light is the great and obsessing enterprise in the three worlds of human evolution; everywhere men rate the light of the sun as essential to healthy living; some idea of the human urge to light can be grasped if you consider the brilliance of the physically engendered light in which we live when night arrives, and compare it with the mode of lighting the streets and homes of the world prior to the discovery of gas, and later of electricity. The light of knowledge, as the reward of educational processes, is the incentive behind all our great schools of learning in every country in the world and is the goal of much of our world organisation; the terminology of light controls even our computation of time. The mystery of electricity is unfolding gradually before our rapt eyes and the electrical nature of man is being slowly proven and will later demonstrate that, throughout the human structure and form, man is composed primarily of light atoms, and that the light in the head (so familiar to esotericists) is no fiction or figment of wishful thinking or of a hallucinated imagination, but is definitely brought about by the junction or fusion of the light inherent in substance itself and the light of the soul.

It will be found that this will be capable of scientific proof. It will also be shown that the soul itself is light, and that the entire Hierarchy is a great centre of light, causing the symbology of light to govern our thinking, our approach to God, and enabling us to understand somewhat the meaning of the words of Christ "I am the Light of the world." These words carry meaning to all true disciples and present them with an analogous goal which they define to themselves as that of finding the light, appropriating the light, and themselves becoming light-bearers. The theme of light runs through all the world Scriptures; the idea of enlightenment conditions all the training given to the youth of the world (limited though the application of this idea may be), and the thought of more light governs all the inchoate yearnings of the human spirit. *The Rays and the Initiations*, pp. 142/3

• [The soul]: In the field of modern psychology we can look for a gradual recognition of the fact of the self. The problem of the psychologists is to comprehend the relationship or the identity of that self with the soul.

It is, however, from the field of science that the greatest help will come. The fact of the soul will eventually be proved through the study of light and of radiation and through a coming evolution in particles of light. Through this imminent development we shall find ourselves seeing more and penetrating deeper into that which we see today. One of the recognised facts in the realm of natural science has been the cyclic change in the fauna and flora of our planet. Animals, plentiful and familiar many thousands of years ago, are now extinct, and by means of their bones we endeavour to reconstruct their forms. Flowers and trees that once covered the surface of our planet have now entirely disappeared and only their fossilized remains are left to indicate to us a vegetation vastly different to that which we now enjoy. Man himself has changed so much that we find it difficult to recognise homo sapiens in the early primitive races of the far distant past. This mutability and obliteration of earlier types is due to a major factor among many. The quality of the light which promotes and nurtures growth, vitality and fertility in the kingdoms of nature has changed several times during the ages, and as it has changed it has produced corresponding mutations in the phenomenal world. From the standpoint of the esotericist, all forms of life on our planet are affected by three types of light substance, and at the present time a fourth type is gradually making its presence felt. These types of light are:

1. The light of the sun.

2. The light in the planet itself—not the reflected light of the sun but its own inherent radiance.

3. A light seeping in (if I may use such a phrase) from the astral plane, a steady and gradual penetration of the "astral light" and its fusion with the other two types of radiance.

4. A light which is beginning to merge itself with the other three types and which comes from that state of matter which we call the mental plane—a light in its turn reflected from the realm of the soul.

An intensification of the light is going on all the time, and this increase in intensity began on the earth at about the time when man discovered the uses of electricity, which discovery was a direct result of this intensification. The electrification of the planet through the widespread use of electricity is one of the things which is inaugurating the new age, and which will aid in bringing about the revelation of the presence of the soul. Before long this intensification will become so great that it will materially assist in the rending of the veil which separates the astral plane from the physical plane; the dividing etheric web will shortly be dissipated, and this will permit a more rapid inflow of the third aspect of light. The light from the astral plane (a starry radiance) and the light of the planet itself will be more closely blended, and the result upon humanity and upon the three other kingdoms in nature cannot be over-emphasized. It will, for one thing, profoundly affect the human eye and make the present sporadic etheric vision a universal asset. It will bring within the radius of our range of contact the infra-red and ultra-violet gamut of colours, and we shall see what at present is hidden. All this will tend to destroy the platform upon which the materialists stand, and to pave the way, first, for the admission of the soul as a sound hypothesis, and secondly, for the demonstration of its existence. We only need more light, in the esoteric sense, in order to see the soul, and that light will shortly be available and we shall understand the meaning of the words, "And in Thy light shall we see light."

This intensification of the light will continue until A.D. 2025, when there will come a cycle of relative stability and of steady shining without much augmentation. In the second decanate of Aquarius these three aspects will again be augmented by increased light from the fourth aspect, that is the light from the soul realm, reaching us via the universal "chitta" or mind stuff. This will flood the world. By that time, however, the soul will be recognised as a fact, and as a consequence of this recognition our entire civilisation will have changed so radically that we cannot today even guess at the form it will take. The next ten years will see a greatly increased merging of the first three forms of light, and those of you who are awake to these issues and happenings will find it interesting to note what is going on.

Esoteric Psychology I, pp. 101/3

• The force of mind. This is the illuminating energy which "lights the way" of an idea or form to be transmitted and received. Forget not that light is subtle substance. Upon a beam of light can the energy of the mind materialise. This is one of the most important statements made in connection with the science of telepathy.

Telepathy and the Etheric Vehicle, p. 26

• "Electric fire, or will-impulse" in conjunction with "fire by friction" produces light or "solar fire." Electric fire is force or energy of some kind, and hence in itself is fundamentally an emanation. "Fire by friction" is substance with the quality of heat as its predominant characteristic; it is latent heat or sensation. Both these ideas, therefore, convey the idea of duality. An emanation must have its originating source, and heat is but the result of friction, and is necessarily dual. Both these concepts involve facts long antedating the solar system, and hidden in the Universal Mind. All that we can scientifically ascertain is the nature of that which is produced by their approximation, and this is solar fire or light. These thoughts may make clear somewhat the significance of the number five, esoterically considered. Electric fire, being an emanation is essentially dual in concept, and so is fire by friction; they together produce solar fire, and thus the esoteric fifth. *A Treatise on Cosmic Fire,* pp. 802/3

• All true esoteric activity produces light and illumination; it results in the inherited light of substance being intensified and qualified by the higher light of the soul—in the case of humanity consciously functioning. It is therefore possible to define esotericism and its activity in terms of light, but I refrain from doing so because of the vagueness and the mystical application hitherto developed by esotericists in past decades. If esotericists would accept, in its simplest form, the pronouncement of modern science that substance and light are synonymous terms, and would recognise also that the light which they can bring to bear on substance (the application of energy to force) is equally substantial in nature, a far more intelligent approach would be made. The esotericist does deal with light in its three aspects, but it is preferable today to attempt a different approach until—through development, trial and experiment—the esotericist knows these triple differentiations in a practical sense and not just theoretically and mystically. We have to live down some of the mistakes of the past. *Education in the New Age*, p. 68

• The esotericist knows that in every atom of his body is to be found a point of light. He knows that the nature of the soul is light. For aeons, he walks by means of the light engendered within his vehicles, by the light within the atomic substance of his body and is, therefore, guided by the light of matter. Later, he discovers the light of the soul. Later still, he learns to fuse and blend soul light and material light. Then he shines forth as a Light bearer, the purified light of matter and the light of the soul being blended and focussed.

Glamour: A World Problem, p. 196

• Individualisation is literally the coming together (out of the darkness of abstraction), of the two factors of Spirit and matter by means of a third factor, the intelligent will, purpose and action of an Entity. By the approximation of these two poles light is produced, a flame shines forth, a sphere of radiant glory is seen which gradually increases the intensity of its light, its heat and its radiance until capacity is reached, or that which we call perfection. We should note and distinguish the words *light, heat and radiance*, which are the distinctive features of all individualised entities from Gods to men.

Man is beginning to arrive partially at the secret of this phenomenon through his ability to produce through scientific knowledge, that which is called electric light and which is used by man for illumination, heat and healing. As more anent this matter is discovered by physical plane students, the whole question of existence and of creative activity will become clearer.

A Treatise on Cosmic Fire, p. 345

• . . . it must be remembered that planetary schemes pass into obscuration and "die out," through the withdrawal in all cases of the positive life and energy and of the electric fire which is the animating principle of every system, scheme, globe, kingdom in nature, and human unit. This produces again in every case the dying out of the "solar radiance," or of the light produced by the commingling of the negative and positive energy. All that is left in every case again is the habitual energy of the substance upon which, and through which, the positive energy has had such a remarkable effect. This negative type of force gradually dissipates, or disperses itself, and seeks the central reservoir of energy. The spheroidal form is thus disintegrated.

A Treatise on Cosmic Fire, pp. 835/6

4. VIBRATION

• *The Law of Vibration*.--This is the law of the first plane, and it governs all the atomic subplanes of each plane. It marks the beginning of the work of the Logos, the first setting in motion of mulaprakriti. On each plane the vibration of the atomic subplane sets in motion the matter of that plane. It is the key measure. We might sum up the significance of this law in the words, "light" or "fire." It is the law of fire; it governs the transmutation of differentiated colours back to their synthesis. It controls the breaking up of the One into the seven, and then the reabsorption back into the One. It is really the basic law of evolution, which necessitates involution. It is analogous to the first movement the Logos made to express Himself through this solar system. He uttered the Sound, a threefold Sound, one sound for each of His three systems, and started a ripple on the ocean of space. The Sound grows in volume as time progresses, and when it has reached its full volume, when it is fully completed, it forms one of the notes in the major cosmic chord. Each note has six subtones, which, with the first, make the seven; the Law of Vibration, therefore, comprises eighteen lesser vibrations and three major, making the twenty-one of our three systems. Two multiplied by nine (2x9), makes the necessary eighteen, which is the key number of our love system. Twenty-seven holds hid the mystery of the third system.

On the path of involution, the seven great Breaths or Sounds drove to the atomic subplane of each plane, and there the basic vibration repeated in its own little world the method of logoic vibration, giving rise itself to six subsidiary breaths. We get the same correspondence here as we did in the matter of the Rays, for we shall find that the lines of vibration are 1-2-4-6. Logically this would be so, for involution is negative, receptive, and corresponds to the feminine pole, just as the abstract rays were 2-4-6. This truth requires meditation, and an attempt to think abstractly; it is linked to the fact that the whole second system is receptive and feminine; it concerns the evolution of consciousness of the psyche.

On the path of evolution this law controls the positive aspect of the process. All is rhythm and movement, and when all that evolves on each plane attains the vibration of the atomic subplane, then the goal is reached. When, therefore, we have achieved the first main vibrations, and have perfected vehicles for all evolutions (not merely the human), of fivefold atomic subplane matter, then we have completed the round of evolution for this system. In the coming system we shall add the next two vibrations that complete the scale, and our Logos will then have completed His building.

A Treatise on Cosmic Fire, pp. 574/5

• *Response to Vibration.* It is always recognised in occult circles that the whole object of human evolution is to enable the Thinker to respond to every contact fully and consciously, and thus to utilise his material sheath, or sheaths, as adequate transmitter of such contact. The most easily studied human thought-form is the one the Ego creates through which to function. He builds his sheaths by the power of thought, and the dense physical body is the best sheath that—at any particular stage of evolution—he can at the time manufacture. The same can be predicated of the solar Logos. He builds by the power of thought a body which can respond to that group of vibrations which are concerned with the cosmic physical plane (the only one we can study). It is not yet adequate, and does not fully express the logoic Thinker.

The vibrations to which the systemic thought-form (Solar System) must respond are many in number, but for our purposes might be enumerated as mainly seven:

1. The vibrations of the cosmic physical plane, viewing it as all the matter of that plane which exists outside the logoic ring-pass-not. It concerns the pranic and akashic fluids and currents.
2. The vibrations of the cosmic astral plane as they affect the physical form of divine manifestation. This involves cosmically the action upon our solar Logos of the emotional quality of other cosmic entities, and concerns the magnetic effect upon Him of their psychic emanation. This, in view of the fact that His dense physical body is not a principle, is of a more potent nature than the first set of vibrations, as is the case also in man's evolution.
3. Vibrations from that which, within the logoic consciousness, is recognised as the logoic Higher Self, or His emanating source. This brings the solar system within the vibratory radius of certain constellations which have a position of profound importance in the general evolution of the system.
4. Vibrations from Sirius via the cosmic mental plane.
5. Vibrations from the seven Rishis of the Great Bear, and primarily from those two Who are the Prototypes of the Lords

of the seventh and fifth Rays. This is a most important point, and finds its microcosmic correspondence in the place which the seventh Ray has in the building of a thoughtform, and the use of the fifth Ray in the work of concretion. All magicians who work with matter and who are occupied with form-building (either consciously or unconsciously) call in these two types of force or energy.

6. Certain very remote vibrations, as yet no more appreciable in the logoic Body than is monadic influence in that of average man, from the ONE ABOUT WHOM NOUGHT MAY BE SAID, that cosmic Existence Who is expressing Himself through seven centres of force, of which our solar system is one.

7. A series of vibrations which will become more potent as our Logos nears that period which is occultly called "Divine Maturity," which emanate from that constellation in the Heavens which embodies His polar opposite. This is a deep mystery and concerns the cosmic marriage of the Logos.

It will be apparent, therefore, how little can as yet be predicated anent the future of the solar system until the vibrations of the sixth and seventh order become more powerful, and their effects can consequently be studied more easily. It is not possible here to do more than indicate the seven types of vibrations to which our solar Logos (functioning in a material body) will in due course of time consciously, and fully, respond. He responds to vibrations of the first, second, third and fourth order quite fully at this time, but as yet (though responding) cannot fully, and consciously, utilise these types of energy. The vibration of the fifth order is recognised by Him, particularly in three of His centres, but is not as yet fully under His control. The other two are sensed, and felt, but so remotely as to be almost outside the range of His consciousness.   *A Treatise on Cosmic Fire,* pp. 552/4

5. COLOUR

• It might here be asked why colour primarily is spoken of as the buddhic manifestation of electricity. We are employing the word "colour" here in its original and basic sense as "that which veils." Colour veils the sevenfold differentiation of logoic manifestation and, from the angle of vision of man in the three worlds, can be seen only in its full significance on the buddhic plane. All fire and electrical display will be seen to embody the seven colours.

A Treatise on Cosmic Fire, p. 321

• The very use of the word "Colour" . . . conveys the idea of concealment. Colour is therefore "that which does conceal." It is simply the objective medium by means of which the inner force transmits itself; it is the reflection upon matter of the type of influence that is emanating from the Logos, and which has penetrated to the densest part of His solar system. We recognise it as colour. The adept knows it as differentiated force, and the initiate of the higher degrees knows it as ultimate light, undifferentiated and undivided.

Letters on Occult Meditation, p. 211

• [Subsidiary aspect of the Law of Attraction]:
The Law of Colour.

To get any comprehension of this law students should remember that colour serves a twofold purpose. It acts as a veil for that which lies behind, and is therefore attracted to the central spark; it demonstrates the attractive quality of the central life.

All colours, therefore, are centres of attraction, are complementary, or are antipathetic to each other, and students who study along these lines can find out the law, and comprehend its working through a realisation of the purpose, the activity, and the relation of colours to or for each other. *A Treatise on Cosmic Fire*, pp. 1171/2

• Every individual vibrates to some particular measure. Those who know and who work clairvoyantly and clairaudiently find that all matter sounds, all matter pulsates, and all matter has its own colour. Each human being can therefore be made to give forth some specific sound; in making that sound he flashes into colour, and the combination of the two is indicative of some measure which is peculiarly his own.

Every unit of the human race is on some one of the seven rays; therefore some one colour predominates, and some one tone sounds forth; infinite are the gradations and many the shades of colour and tone. Each ray has its subsidiary rays which it dominates, acting as the synthetic ray. These seven rays are linked with the colours of the spectrum. There are the rays of red, blue, yellow, orange, green and violet. There is the ray that synthesises them all, that of indigo. There are the three major rays—red, blue and yellow—and the four subsidiary colours which, in the evolving Monad, find their correspondence in the spiritual Triad and the lower quaternary. The Logos of our system is concentrating on the love or blue aspect. This—as the synthesis—manifests as indigo. *Esoteric Psychology* I, pp. 126/7

a. *Forecasts on Colour, Sound and Vibration*

• Forecasts anent the future.

1. The phraseology of the medical schools will more and more become based on vibration and be expressed in terms of sound and colour.

2. The religious teaching of the world and the inculcation of virtue will be likewise imparted in terms of colour. People will eventually be grouped under their ray-colour, and this will be possible as the human race develops the faculty of seeing auras. The number of clairvoyants is already greater than is realised, owing to the reticence of the true psychic.

3. The science of numbers, being in reality the science of colour and sound, will also somewhat change its phraseology and colours will eventually supersede figures.

4. The laws that govern the erection of large buildings and the handling of great weights will some day be understood in terms of sound. The cycle returns, and in the days to come will be seen the reappearance of the faculty of the Lemurians and early Atlanteans to raise great masses,—this time on a higher turn of the spiral. Mental

comprehension of the method will be developed. They were raised through the ability of the early builders to create a vacuum through sound, and to utilise it for their own purposes.

5. Destruction, it will be shewn, can be brought about by the manipulation of certain colours, and by the employment of united sound. In this way terrific effects will be achieved. Colour can destroy just as it can heal; sound can disrupt just as it can bring about cohesion; in these two thoughts lie hid the next step ahead for the science of the immediate future. The laws of vibration are going to be widely studied and comprehended and the use of this knowledge of vibration on the physical planes will bring about many interesting developments. They will be partially an outgrowth of the study of the war and its effect, psychological and otherwise. More was effected by the sound of the great guns, for instance, than by the impact of the projectile on the physical plane. These effects are as yet practically unrecognised, and are largely etheric and astral.

6. Music will be largely employed in construction, and in one hundred years from now it will be a feature in certain work of a constructive nature. This sounds to you utterly impossible, but it will simply be the utilisation of ordered sound to achieve certain ends.

You will ask, what place has all this in a series of letters on meditation? Simply this:—that the method employed in the utilisation of colour and sound in healing, in promoting spiritual growth, and in exoteric construction on the physical plane, will be based on the laws that govern the mental body, and will be forms of meditation. Only as the race develops the dynamic powers and attributes of thought— which powers are the product of meditation, rightly pursued—will the capacity to make use of the laws of vibration be objectively possible. Think not that only the religious devotee or mystic, or the man imbued with what we call higher teaching, is the exponent of the powers attained by meditation. All great capitalists, and the supreme heads of finance, or organised business, are the exponents of similar powers. They are personifications of one-pointed adherence to one line of thought, and their evolution parallels that of the mystic and the occultist. I seek most strongly to emphasise this fact. They are the ones who meditate along the line of the Mahachohan, or the Lord of Civilisation or Culture. Supreme concentrated attention to the matter in hand makes them what they are, and in many respects they attain greater results than many a student of meditation. All they need to do is to transmute the motive underlying their work, and their achievement will then outrun that of other students. They will approach a point of synthesis, and the Probationary Path will then be trodden.

The Law of Vibration will gradually, therefore, be more and more understood, and be seen to govern action in all of the three departments of the Manu, the World Teacher and the Mahachohan. It will find its basic expression and its familiar terminology in those of colour and sound. Emotional disorder will be regarded as discordant sound; mental lethargy will be expressed in terms of low vibration, and physical disease will be numerically considered. All constructive work will eventually be expressed in terms of numbers, by colours, and through sound.

This suffices on this matter and at this juncture I have naught further to communicate. The subject is abstruse and difficult, and only by patient brooding will the darkness lighten. Only when the ray of the intuition strikes athwart the pall of darkness (which pall is the ignorance that hides all knowledge) will the forms that veil the subjective life be irradiated and known. Only when the light of reason is dimmed by the radiant sun of wisdom will all things be seen in their just proportions, and will the forms assume their exact colours, and their numerical vibration be known.

Letters on Occult Meditation, pp. 248/52

6. MAGNETISM

• Magnetism, and the capacity to show love, are occultly synonymous. *A Treatise on Cosmic Fire,* p. 576

• It might here be useful to point out that *magnetism* is the effect of the divine ray (2nd Ray) in manifestation in the same sense that electricity is the manifested effect of the primordial ray (3rd Ray) of active intelligence. It would be well to ponder on this for it holds hid a mystery. *A Treatise on Cosmic Fire*, p. 44

• *The law of Magnetic Control.* —This law is the basic law controlling the Spiritual Triad. Through this law, the force of evolution drives the Ego to progress through the cycle of reincarnation back to union with his kind. Through separation he finds himself, and then— driven by the indwelling buddhic or Christ principle—transcends himself, and finds himself again in all selves. This law holds the evolving lower self in a coherent form. It controls the Ego in the causal body, in the same way that the Logos controls the Monad on the second plane. It is the law of the buddhic plane; the Master is one Who can function on the buddhic levels, and Who has magnetic control in the three worlds. The lower is always controlled from above, and the effect the buddhic levels have on the three lower is paramount, though that is scarcely yet conceded by our thinkers. It is the Law of Love, in the three worlds, that holds all together, and that draws all upward. It is the demonstration, in the Triad, of the Law of Attraction.

On the path of involution this law works with the permanent atoms in the causal body. It is the buddhic principle, and its relation with the lower permanent atom of the Triad is the mainspring of the life of the Ego. On the path of descent it has much to do with the placing of the permanent atoms, but this matter is very abstruse, and the time has not yet come for further elucidation. At the third outpouring, (in which the fourth kingdom, the human, was formed), it was this Law of Magnetic Control that effected the juncture of astro-animal man, and the descending Monad, using the spark of mind as the method of at-one-ment. Again we can see how it works. The monadic plane, the buddhic plane, and the astral plane are all three closely allied, and we find there the line of least resistance.

A Treatise on Cosmic Fire, pp. 583/4

• The fourth law, Magnetic Control, for instance, holds sway on the fourth subplane of each plane, in the fourth round, and in the fourth root-race specially. We shall then have the correspondence as follows:

- 4th Law . . . . Magnetic Control.
- 4th Ray . . . . Harmony or beauty.
- 4th Plane . . . . The buddhic.
- 4th Subplane . . . . . . Buddhic Magnetic Control.
- 4th Round . . . . Dense Physical Magnetism, controlling sex manifestation on the physical plane, and inspired by astral desire, the reflection of the buddhic.
- 4th Root-Race. . . . . . The Atlantean, in which the above qualities specially demonstrated.

A Treatise on Cosmic Fire, pp. 573/4

• The Law of Magnetic Impulse

. . . It would be well to remember that we are not considering here that aspect of the second ray which is peculiarly concerned with form, and which constitutes the cohering, magnetic agent in any form, whether atom, man or solar system. We are not here concerned with the relation between forms, even though due (as is essentially the case) to second ray energy. Nor are we occupied with considering the relation of soul to form, either that of the One Soul to the many forms, or of an individualised soul to its imprisoning form. The laws we are considering are concerned entirely with inter-soul relationship, and with the synthesis underlying the forms. They govern the conscious contact existing between the many aspects of the One Soul. I have expressed this phrase with care.

This Law of Magnetic Impulse governs the relation, the interplay, the intercourse, and the interpenetration between the seven groups of souls on the higher levels of the mental plane which constitute the first of the major form differentiations. These we can only study intelligently from the angle of the seven ray groups, as they compose the spiritual aspect of the human family. This law governs also the relationships between souls, who, whilst in manifestation through form, are en rapport with each other. It is a law, therefore, which concerns the inter-relation of all souls within the periphery of what the Christians call "the Kingdom of God." Through a right understanding of this law, the man arrives at a knowledge of his subjective life; he can wield power subjectively, and thus work consciously in form and with form, yet holding his polarisation and his consciousness in another dimension, and functioning actively behind the scenes. This law concerns those inner esoteric activities which are not primarily related to form life.

This law is of major importance because of the fact that Deity itself is on the second ray; because this is a second ray solar system, and therefore all rays and the varying states or groupings of consciousness, and all forms, in or out of physical manifestation, are coloured and dominated by this ray, and therefore again finally controlled by this law. The Law of Magnetic Impulse is in the soul realm what the Law of Attraction is in the world of phenomena. It is, in reality, the subjective aspect of that Law. It is the Law of Attraction as it functions in the kingdom of souls, but because it is functioning on those levels where the "great heresy of separateness" is not to be found, it is difficult for us—with our active, discriminating minds—to understand its implications and its significance. This Law governs the soul realm, to it the Solar Angels respond, and under its stimulation, the egoic lotuses unfold. It could perhaps be best understood if it is considered as—

a. The impulsive interplay between souls in form and out of form.
b. The basis of egoic recognition.
c. The factor which produces reorientation in the three worlds.
d. The cause of the magnetic rapport between a Master and His group, or a Master and His disciple.

It has an occult name, and we call it "the Law of Polar Union." Yet when I say to you that this implies the binding of the pairs of opposites, the fusion of the dualities, and the marriage of souls, I have uttered meaningless words, or words which—at the best—embody an ideal which is so closely tied up with material things in the mind of the aspirant, and so connected with the processes of detachment (at which disciples work so strenuously!) that I despair of presenting the truth as it concerns souls and soul relationships.

This law governs also the relation of the soul of a group to the soul of other groups. It governs the interplay, vital but unrealised yet as a potency, between the soul of the fourth kingdom in nature, the human, and the soul of the three subhuman kingdoms, and likewise that of the three superhuman kingdoms. Owing to the major part which humanity has to play in the great scheme or Plan of God, this is the law which will be the determining law of the race. This will not, however, be the case until the majority of human beings understand something of what it means to function as a soul. Then, under obedience to this law, humanity will act as a transmitter of light, energy and spiritual potency to the subhuman kingdoms, and will constitute a channel of communication between "that which is above and that which is below." Such is the high destiny before the race.

Just as certain human beings have, through meditation, discipline and service, most definitely made a contact with their own souls, and can therefore become channels for soul expression, and mediums for the distribution into the world of soul energy, so men and women, who are oriented to soul living in their aggregate, form a group of souls, en rapport with the source of spiritual supply. They have, as a group, and from the angle of the Hierarchy, established a contact and are "in touch" with the world of spiritual realities. Just as the individual disciple stabilises this contact and learns to make a rapid alignment and then, and only then, can come into touch with the Master of his group and intelligently respond to the Plan, so does this group of aligned souls come into contact with certain greater Lives and Forces of Light, such as the Christ and the Buddha. The aggregated aspiration, consecration and intelligent devotion of the group carries the individuals of which it is composed to greater heights than would be possible alone. The group stimulation and the united effort sweep the entire group to an intensity of realisation that would otherwise be impossible. Just as the Law of Attraction, working on the physical plane, brought them together as men and women into one group effort, so the Law of Magnetic Impulse can begin to control them when, again as a group and only as a group, they unitedly constitute themselves channels for service in pure self-forgetfulness.

This thought embodies the opportunity immediately before all groups of aspirants and allied men of good will in the world today. If they work together as a group of souls, they can accomplish much. This thought illustrates also the significance of this law which does produce polar union. What is needed to be grasped is that in this work, there is no personal ambition implied, even of a spiritual nature and no personal union sought. This is not the mystical union of the scriptures or of the mystical tradition. It is not alignment and union with a Master's group, or fusion with one's inner band of pledged disciples, nor even with one's own Ray life. All these factors constitute preliminary implications and are of an individual application. Upon this sentence I ask you to ponder. This union is a greater and more vital thing because it is a group union. *Esoteric Psychology* II, pp. 109/13

• *Path II. The Path of Magnetic Work.*

In considering this Path students must bear in mind that they are dealing with that Path which of all the seven expresses most fully the effects of the Law of Attraction. It will be remembered by those who have carefully read this *Treatise* that this law is the expression of the spiritual will which produces the manifestation of the Son (Sun). Magnetism—physical, attractive and dynamic—is the expression of the law in the three worlds as far as the human unit is concerned. It will be apparent, therefore, that the adept who passes upon this Path is dealing with that reality which is the basis of all *coherency* in nature, and with that essence which through the force of its own innate quality produces the attractive energy which brings together the pairs of opposites; it is the force which is responsible for the interplay of electrical phenomena of every kind. . . .

Those who do the work of wielding forces or electrical magnetism for the use of the Great Ones on all planes pass to this Path. They wield the elemental formative energy, manipulating matter of every density and vibration. Great waves of ideas and surging currents of public opinion on astral levels as well as on the higher levels where work the Great Ones, are manipulated by them. A large num-

ber of fifth Ray people, those who have the Ray of Concrete Knowledge for their monadic ray, pass to this line of endeavour.
A Treatise on Cosmic Fire, pp. 1247/8

• The entire problem of magnetism is closely connected with the problem of sex. In the occult study of the dissemination of the seed life and the germs of the vegetable kingdom, and in the understanding of the part played therein by those miraculously developed organisms,— the ants and bees—and later in the investigation of the work of the etheric builders, the elves and fairies, by those with awakened vision will come a new light upon sex and upon the function it serves in the interrelation of lives and the creation of forms. With this aspect of this deeply esoteric truth I cannot here deal, for it is the effect of activity in the solar lives of the solar system, and with these we cannot concern ourselves. *Esoteric Psychology* I, pp. 244/5

• Where the disciple is concerned, release from the constant consideration of personal circumstances and problems leads inevitably to a clear mental release; this then provides those areas of free mental perception which make the higher sensitivity possible. Gradually, as the disciple acquires true freedom of thought and the power to be receptive to the impression of the abstract mind, he creates for himself a reservoir of thought which becomes available at need for the helping of other people and for the necessities of his growing world service. Later, he becomes sensitive to impression from the Hierarchy. This is at first purely ashramic, but is later transformed into total hierarchical impression by the time the disciple is a Master; the Plan is then the dynamic substance providing the content of the reservoir of thought upon which he can draw. This is a statement of unique and unusual importance. . . .

The essential point to be grasped is that sensitivity to impression is a normal and natural unfoldment, paralleling spiritual development. I gave you a clue to the entire process when I said that

"Sensitivity to impression involves the engendering of a magnetic aura upon which the highest impressions can play."

I would have you give the deepest consideration to these words. As the disciple begins to demonstrate soul quality, and the second divine aspect takes possession of him and controls and colours his entire life, automatically the higher sensitivity is developed; he becomes a magnet for spiritual ideas and concepts; he attracts into his field of consciousness the outline, and later the details, of the hierarchical Plan; he becomes aware eventually of the planetary Purpose; all these impressions are not things which he must seek out and learn laboriously to ascertain, to hold and seize upon. They drop into his field of consciousness because he has created a magnetic aura which invokes them and brings them "into his mind". This magnetic aura begins to form itself from the first moment he makes a contact with his soul; it deepens and grows as those contacts increase in frequency and become eventually an habitual state of consciousness; then, at will and at all times, he is en rapport with his soul, the second divine aspect.

Telepathy and the Etheric Vehicle, pp. 94/6

7. POLARITY

• At the fourth initiation another of the great secrets is revealed to him. It is called "the mystery of polarity," and the clue to the significance of sex in every department of nature on all the planes is given to him. It is not possible to say much along these lines. All that can be done is to enumerate some of the subjects to which it gives the clue, adding to this the information that in our planetary scheme, owing to the point in evolution of our own Planetary Logos, this secret is the most vital. Our Planetary Logos is at the stage wherein He is consciously seeking the at-one-ment with his polar opposite, another Planetary Logos. The subjects on which this secret throws a flood of light are:—

a. Sex on the physical plane. It gives us a key to the mystery of the separation of the sexes in Lemurian days.
b. The balancing of forces in all departments of nature.
c. The clue as to which Scheme forms with ours a duality.
d. The true name of our Planetary Logos and His relation to the

Solar Logos.
- e. "The Marriage of the Lamb" and the problem of the heavenly bride. A clue to this lies in the solar system of S.... which must be read astrologically.
- f. The mystery of the Gemini, and the connection of our particular Planetary Logos with that constellation.

On a lesser scale, and in relation to the microcosm, the following subjects are illuminated when the initiate receives the second great secret, or the fourth which includes the earlier lesser ones:—

- g. The processes of at-one-ment in the different kingdoms of nature. The bridging between the kingdoms is shown him, and he sees the unity of the scheme.
- h. The method of egoic at-one-ment is seen clearly revealed, and the antahkarana is shown in its real nature, and having been thus revealed, is dispensed with.
- i. The essential unity existing between the Ego and the personality is seen.
- j. The relation of the two evolutions, human and deva, is no longer a mystery, but their position in the body of the Heavenly Man is seen to be a fact.

One could go on emphasising the multiplicity of matters which the mystery of polarity, when revealed, makes clear to the initiate, but the above suffices. This secret concerns primarily the Vishnu, or second aspect. It sums up in one short phrase the totality of knowledge gained in the Hall of Wisdom, as the earlier secrets summed up the totality achieved in the Hall of Learning. It deals with consciousness and its development by and through the matter aspect. It concerns literally the unification of the self and the not-self till they are verily and indeed one. *Initiation, Human and Solar,* pp 172/4

• This question of the electrical polarity of the centres is one of real difficulty, and little can be communicated on the matter. It may be safely pointed out, however, that the generative organs are the negative pole to the throat centre as is the solar plexus to the heart. The order of the development of the centres, the ray-type and colour, coupled to the fact that during certain stages of the evolutionary process different centres (such as the base of the spine) are positive to all the others, not even excluding the head centre, leads to the vast complexity of the subject. Likewise certain of the planetary schemes are positive and others negative; three of the schemes are dual, both negative and positive. The same can be predicated anent a solar system, and, curiously enough, anent the planes themselves. For instance, in connection with the earth scheme we have a positive polarity of a temporary nature based on the type of incarnation our particular Heavenly Man is undergoing on our planet. By this is meant that there are masculine and feminine incarnations undergone by Heavenly Men as by men, considering the entire subject from the angle of electrical polarity and not from sex as understood in connection with the physical body.

Venus is negatively polarised, and hence it became possible for a mysterious absorption by the Earth of Venusian force. Again in this connection the question of sex may serve to elucidate. The karmic tie between the two Heavenly Men—one in a positive incarnation and the other in a negative—caused the working out of an old debt and a planetary alliance. *Light* flashed forth in Lemurian days in a number of great groups of the human family when these two opposite poles made electrical connection. It needed the joint work of the two Heavenly Men, working on buddhic levels (the fourth cosmic ether) to bring about the blazing forth of the light of manas in the causal groups on the fifth cosmic gaseous plane, the mental plane of the solar system. . . .

It must ever be borne in mind that each plane and each subplane which is numerically allied, is embodying the same type of force, and is consequently of the same polarity.

A Treatise on Cosmic Fire, pp. 323/4

• *Relation.* Another outstanding feature that is the result of our studies is that of relation. The realisation of this in future years will lead to the study of the different polarities of the different spheres (from a planetary scheme to an atom) within the solar ring-pass-not, and of the relation existing between:

a. A scheme and the totality of schemes.
b. Scheme and scheme.
c. Chain and chain.
d. Globe and globe.
e. Group and group.
f. Subdivision and subdivision.
g. Unit and unit.
h. Cell and cell.

The interrelation of all these factors and their profound interdependence is one of the most important points for us to grasp; though this whole relation is governed by the law of Attraction and Repulsion, and therefore comes more under what we call the second aspect, yet self-consciousness itself is the result of the manasic principle, and the close co-operation between these two factors of mind and love-wisdom, or the two laws of Attraction and Synthesis, must ever be carefully remembered. *A Treatise on Cosmic Fire,* pp. 410/1

8. GRAVITATION

• [Subsidiary aspect of the Law of Attraction]:
The Law of Gravitation.

This law is for the non-occult student the most puzzling and confusing of all the laws. It shows itself in one aspect as the power, and the stronger urge that a more vital life may have upon the lesser, such as the power of the spirit of the Earth (the planetary Entity, not the planetary Logos) to hold all physical forms to itself and prevent their "scattering." This is due to the heavier vibration, the greater accumulative force, and the aggregated tamasic lives of the body of the planetary Entity. This force works upon the negative, or lowest, aspect of all physical forms. The Law of Gravitation shows itself also in the response of the soul of all things to the greater Soul in which the lesser finds itself. This law, therefore, affects the two lowest forms of divine life, but not the highest. It emanates in the first instance from the physical sun and the heart of the Sun. The final synthesising forces which might be regarded as forms of spiritual gravitational activity are, nevertheless, not so, but are due to the working of another law, emanating from the central spiritual Sun. The one is purely systemic, the other a cosmic law. *A Treatise on Cosmic Fire*, p. 1172

• One of those maxims I can here give anent supply and demand. It is only as a skilful use is made of the supply for the needs of the worker and the work (I choose these words each one with deliberation) that that supply continues to pour in. The secret is: use, demand, take. Only as the door is unlocked by the law of demand is another and higher door unlocked permitting supply. The law of gravitation holds hid the secret. Think this out.

Letters on Occult Meditation, pp. 204/5

• The Law of Cause and Effect is not to be understood as we now interpret it. There is, to illustrate, a law called the Law of Gravitation, which has long imposed itself upon the minds of men; such a law exists, but it is only an aspect of a greater law, and its power can be, as we know, relatively offset, for each time that we see an aeroplane soaring overhead, we see a demonstration of the offsetting of this law by mechanical means, symbolising the ease with which it can be surmounted by human beings. If they could but realise it, they are learning the ancient technique of which the power to levitate is one of the easiest and simplest initial exercises. *Esoteric Healing*, p. 20

• The great Laws can be transcended and frequently have been in the past, and increasingly will be in the future. The Law of Gravitation is frequently offset and daily transcended when an aeroplane is in flight. The energy of faith can set in motion superior energies which can negate or retard disease. The whole subject of faith, and its vital significance and potency, is as little understood as is the Law of Karma. This is a tremendous subject, and I cannot further enlarge upon it. But I have said enough to offer you food for thought.

Esoteric Healing, p. 350

9. HEAT

• *There exists in the Sun, in the planet, in man, and in the atom, a central point of heat, or (if I might use so limiting and inappropriate a term) a central cavern of fire, or nucleus of heat, and this central nucleus reaches the bounds of its sphere of influence, its ring-pass-not by means of a threefold channel.*

A Treatise on Cosmic Fire, p. 58

• Let us now briefly recognise certain facts regarding fire in matter and let us take them in order, leaving time to elucidate their significance. First we might say that the internal fire being both latent and active, shows itself as the synthesis of the acknowledged fires of the system, and demonstrates, for instance, as solar radiation and inner planetary combustion. This subject has been somewhat covered by science, and is hidden in the mystery of physical plane electricity, which is an expression of the active internal fires of the system and of the planet just as inner combustion is an expression of the latent internal fires. These latter fires are to be found in the interior of each globe, and are the basis of all objective physical life.

Secondly, we might note that the internal fires are the basis of life in the lower three kingdoms of nature, and in the fourth or human kingdom in connection with the two lower vehicles. The Fire of Mind, when blended with the internal fires, is the basis of life in the fourth kingdom, and united they control (partially now and later entirely) the lower threefold man or the personality; this control lasts up to the time of the first Initiation.

The fire of Spirit finally, when blended with the two other fires (which blending commences in man at the first initiation), forms a basis of spiritual life or existence. As evolution proceeds in the fifth or spiritual kingdom, these three fires blaze forth simultaneously, producing perfected consciousness. This blaze results in the final purification of matter and its consequent adequacy; at the close of manifestation it brings about eventually the destruction of the form and its dissolution, and the termination of existence as understood on the lower planes. In terms of Buddhistic theology it produces annihilation; this involves, not loss of identity, but the cessation of objectivity and the escape of Spirit, plus mind, to its cosmic centre. It has its analogy in the initiation at which the adept stands free from the limitations of matter in the three worlds.

The internal fires of the system, of the planet, and of man are threefold:

1. Interior fire at the centre of the sphere, those inner furnaces which produce warmth. This is latent fire.
2. Radiatory fire. This type of fire might be expressed in terms of physical plane electricity, of light rays, and of etheric energy. This is active fire.
3. Essential fire, or the fire elementals who are themselves the essence of fire. They are mainly divided into two groups:
    a. Fire devas or evolutionary entities.
    b. Fire elementals or involutionary entities.
    Later we will elaborate on this when we consider the Fire of Mind and deal with the nature of the thought elementals. All these elementals and devas are under the control of the fire Lord, Agni. . . .

We might here point out, however, that our first two statements concerning the internal fires, express the effect that the fire entities have upon their environment. Heat and radiation are other terms which might be applied in this sense. Each of these effects produces a different class of phenomena. Latent fire causes the active growth of that in which it is embedded and causes that upward pushing which brings into manifestation all that is found in the kingdoms of nature. Radiatory fire causes the continued growth of that which has progressed, under the influence of latent fire, to a point receptive of the radiatory. *A Treatise on Cosmic Fire*, pp. 51/3

• *Frictional activity*. This, as is apparent from the words, deals with the "Fire by friction" aspect of substance, and therefore with the lowest aspect of the energy of the mental sheath. The force of the Life within the sheath manifests in the attractive and repulsive action of the individual atoms, and this constant and ceaseless interplay results in the "occult heat" of the body, and its increased radiation. It is one of the factors also which produces the gradual building in of new atoms of substance (ever of a better and more adequate quality) and the expulsion of that which fails to suffice as a medium for intelligent expression.   *A Treatise on Cosmic Fire,* p. 1106

- THE THREE CHANNELS FOR THE FIRE

From the very use of the term "sheath" it will be noted that we are considering those fires which manifest through the medium of those externalities, of those veils of substance which hide and conceal the inner Reality. We shall not here take up the subject of the sheaths on the higher planes, but simply deal with the fires that animate the three lower vehicles,—the physical body in its two divisions (etheric and dense), the emotional or astral body, and the mental sheath. It is frequently overlooked by the casual student that both the astral and the mental bodies are material, and just as material in their own way, as is the dense physical body, and also that the substance of which they are composed is animated by a triple fire, as is the physical.

In the physical body we have the fires of the lower nature (the animal plane) centralised at the base of the spine. They are situated at a spot which stands in relation to the physical body as the physical sun to the solar system. This central point of heat radiates in all directions, using the spinal column as its main artery, but working in close connection with certain central ganglia, wherever located, and having a special association with the spleen.

In the etheric body, which is an exact replica of its denser counterpart, we have the organ of active or radiatory fire, and, as is well known, the vehicle of prana. Its function is to store up the rays of radiatory light and heat which are secured from the sun, and to transmit them, via the spleen, to all parts of the physical body. Hence in the future it will come to be recognised that the spine and the spleen are of

PHYSICAL FORCES OF EVOLUTION

the utmost importance to the physical well-being of man, and that when the spinal column is duly adjusted and aligned, and when the spleen is freed from congestion and in a healthy condition, there will be little trouble in the dense physical body. When the physical furnace burns brightly and when the fuel of the body (pranic rays) is adequately assimilated, the human frame will function as desired.

The subject of the blending of these two fires, which is complete in a normal and healthy person, should engross the attention of the modern physician. He will then concern himself with the removal of nerve congestion or material congestion, so as to leave a free channel for the inner warmth. This blending, which is now a natural and usual growth in every human being, was one of the signs of attainment or of initiation in an earlier solar system. Just as initiation and liberation are marked in this solar system by the blending of the fires of the body, of the mind and of the Spirit, so in an earlier cycle attainment was marked by the blending of the latent fires of matter with the radiatory or active fires, and then their union with the fires of mind. In the earlier period the effects in manifestation of the divine Flame were so remote and deeply hidden as to be scarcely recognisable, though dimly there. Its correspondence can be seen in the animal kingdom, in which instinct holds the intuition in latency, and the Spirit dimly overshadows. Yet all is part of a divine whole.

A Treatise on Cosmic Fire, pp. 55/8
See also: 'The Internal Fire of the Sheaths',
A Treatise on Cosmic Fire, pp. 55/69

10. MOTION

• *The Third Logos.* The third Logos, or Brahma, is characterised by active intelligence; His mode of action is that which we call *rotary*, or that measured revolution of the matter of the system, first as a grand totality, setting in movement the material circumscribed by the entire ring-pass-not, and secondly differentiating it, according to seven vibratory rates or measures into the seven planes. On each of these planes the process is pursued, and the matter of any plane within the plane ring-pass-not shows first as a totality and then as a sevenfold differentiation. This differentiation of matter is brought about by rotary motion, and is controlled by the *Law of Economy* (one of the cosmic laws) with which we will deal later, only pausing here to say that this Law of Economy might be considered as the controlling factor in the life of the third Logos. Therefore:

a. His *goal* is the perfect blending of Spirit and matter.

b. His *function* is the manipulation of prakriti, or matter, so as to make it fit, or equal to, the demands and needs of the Spirit.

c. His *mode of action* is rotary, or, by the revolution of matter, to increase activity and thereby make the material more pliable. . . .

The Second Logos. The second Logos. Vishnu, the divine Wisdom Ray, the great principle of Buddhi seeking to blend with the principle of Intelligence, is characterised by Love. His motion is that which we might term *spiral cyclic*. Availing Himself of the rotary motion of all atoms, He adds to that His own form of motion or of spiralling periodical movement, and by circulation along an orbit or spheroidal path (which circles around a central focal point in an ever ascending spiral) two results are brought about:

a. He gathers the atoms into forms.

b. By means of these forms He gains the needed contact, and develops full consciousness on the five planes of human development, gradually rarefying and refining the forms as the Spirit of Love or the Flame Divine spirals ever onward towards its goal, that goal which is also the source from which it came. . . .

The First Logos. The first Logos is the Ray of Cosmic Will. His mode of action is a literal driving forward of the solar ring-pass-not through space, and until the end of this mahamanvantara or day of Brahma (the logoic cycle) we shall not be able to conceive of the first aspect of will or power as it really is. We know it now as the will to exist, manifesting through the *matter of the forms* (the Primordial Ray and the Divine Ray), and we know it as that which in some occult manner links the system up with its cosmic centre. In a manner inconceivable to us the first Logos brings in the influence of other constellations. When this first aspect is better understood (in the next mahamanvantara) the work of the seven Rishis of the Great Bear, and the supreme influence of Sirius will be comprehended; in this present manifestation of the Son, or of the Vishnu aspect, we are concerned more closely with the Pleiades and their influence via the Sun, and, in relation to our planet, via Venus.

This subject of the first Logos, manifesting only in connection with the other two in the system, is a profound mystery, which is not fully understood by even those who have taken the sixth Initiation.

A Treatise on Cosmic Fire, pp. 142/3, 145/6

a. Rotary

• We will now endeavor to confine ourselves strictly to the subject of fire in matter, and its active effect upon the sheaths of which it is the animating factor, and upon the centers which come primarily under its control.

As we have been told, and as is generally recognised, the effect of heat in matter is to produce that activity which we call rotary, or the revolution of the spheres. Some of the ancient books, and among them a few that are not yet accessible in the occident, have taught that the entire vault of heaven is a vast sphere, revolving slowly like a stupendous wheel, and carrying with it, in its revolution, the entire number of constellations and of universes contained within it. This is a statement unverifiable by the finite mind of man at his present stage, and with his present scientific accessories, but (like all occult statements) it contains within it the seed of thought, the germ of truths, and the clue to the mystery of the universe. Suffice it here to say, that the rotation of the spheres within the solar periphery is a recognized occult fact; and indications are available to prove that science itself likewise formulates the hypothesis that the solar ring-pass-not similarly rotates in its appointed place among the constellations. But at this juncture we will not deal with this angle of the subject, but will study the rotary action of the spheres of the system, and of its content—all the lesser spheres of every degree—remembering ever to keep the distinction clearly in mind that we are dealing now simply with the inherent characteristic of matter itself, and not with matter in co-operation with its opposite, Spirit, which co-operation brings about spiral-cyclic movement.   *A Treatise on Cosmic Fire*, pp. 151/2

• THE QUALITIES OF ROTARY MOTION

Every rotating sphere of matter is characterised by the three qualities, of inertia, mobility and rhythm.

1. *Inertia.* This characterises every atom at the dawn of manifestation, at the beginning of a solar cycle or mahamanvantara (or one hundred years of Brahma), at the commencement of a chain, of a globe, or of any spheroidal form whatsoever without exception. This statement, therefore, includes the totality of manifesting forms within the solar system. . . .

2. *Mobility.* The inherent fires of matter produce rotary movement. Eventually this rotation results in radiation. The radiation of matter, the result of its dual heat, produces necessarily an effect upon other atoms in its environment (it matters not whether that environment is cosmic space, systemic space, or the periphery of the physical body of a man), and this interaction and interplay causes repulsion and attraction according to the polarity of the cosmic, systemic or physical atom. Eventually this produces coherence of form; bodies, or aggregates of atoms come into being or manifestation, and persist for the length of their greater or lesser cycles until the third quality is brought into definite recognition.

3. *Rhythm*, or the attainment of the point of perfect balance and of equilibrium. This point of perfect balance then produces certain specific effects which might be enumerated and pondered upon, even if to our finite minds they may seem paradoxical and contradictory.

The limitation lies with us and with the use of words, and not in any real inaccuracy. These effects are:

a. The disintegration of form,
b. The liberation of the essence which the form confines,
c. The separations of Spirit and matter,

PHYSICAL FORCES OF EVOLUTION

 d. The end of a cycle, whether planetary, human or solar,
 e. The production of obscuration, and the end of objectivity or manifestation,
 f. The reabsorption of the essence, and the merging again of differentiated matter with the root of matter,
 g. The end of time and space as we understand it,
 h. The unification of the three Fires and the bringing about of spontaneous combustion, if one might so express it,
 i. The synthetic activity of matter in the three types of movement,—rotary, spiralling-cyclic and onward progression,—which unified movement will be produced by the interaction of the fires of matter, of mind and of Spirit upon each other.

When the point of rhythm or balance is reached in a solar system, in a plane, in a ray, in a causal body, and in the physical body, then the occupier of the form is loosed from prison; he can withdraw to his originating source, and is liberated from the sheath which has hitherto acted as a prison; and he can escape from an environment which he has utilised for the gaining of experience and as a battle ground between the pairs of opposites. The sheath or form of whatever kind then automatically disintegrates.

A Treatise on Cosmic Fire, pp. 157/9
See also: 'Motion on the Physical and Astral Planes',
A Treatise on Cosmic Fire, pp. 141/214

b. Spiral-Cyclic

• RESULTS OF ITS ACTIVITY

These results can be studied in four ways, considering each as a subsidiary Law of the basic Law of Attraction and Repulsion. All motion is the result literally of the impact, or intercourse, between atoms, and there is no atom anywhere which escapes this force. In the case of rotary motion, which governs the activity of the atom of substance, the impulse emanates from within the ring-pass-not, and is produced by the impact of the positive charge upon the negative charges. This is true of all atoms, cosmic, solar, individual, chemical, and so forth.

When, however, the effect of the rotation of the atom is so strong that it begins to affect other atoms outside its individual ring-pass-not, another influence begins to make itself felt, which draws together, or dissipates, those contacting coalescing atoms. Thus forms are built under the impulse of aggregated forces of some one kind, and these forms in turn produce effects on other cohering atomic forms, until the rhythm is built up, and a vibration instigated which is a continuation of the rotary motion of the individual atoms, and the modification produced on them by their group activity. This causes progression and simultaneous rotation. The movement forward is modified considerably by the internal atomic activity, and this it is which causes that motion we call spiral cyclic. It demonstrates in all forms as a tendency to repeat, owing to the backward pull of the rotating atoms, and yet is offset by the strong progressive impulse of the form activity.
A Treatise on Cosmic Fire, pp. 1039/40

• One of the primary things the occult student should remember when considering the nature of spiral-cyclic activity, is that it has two effects.

First, it is an attractive force, gathering the rotating atoms of matter into definite types and forms, and holding them there as long as necessity demands.

Secondly, it is itself gradually dominated by another and higher vibration, and through its spiralling progress through matter it sweeps those forms systematically nearer and nearer to another and stronger point of energy.

These effects are to be seen clearly demonstrated in man's evolution in the approach he makes uniformly through the cycles to the centre of the spiral-cyclic energy, and subsequently to the still more impressive point, that of his "Father in Heaven." The Angel first attracts animal man; cyclically He actuates the material sheaths, thus giving them coherence, and ever swings them into closer relation to himself. Later, as the momentum is increased, the man is swung more definitely into relation with the monadic aspect, until that higher rhythm is imposed upon him. This is equally true of a planetary

PHYSICAL FORCES OF EVOLUTION

Logos, and of a solar Logos.

A Treatise on Cosmic Fire, pp. 1034/5
See also four laws of Spiral-Cyclic activity:
A Treatise on Cosmic Fire, pp. 1039/83

c. Forward Progressive

• . . . the first Logos controls the cosmic entities or extra-systemic beings; the second Logos controls the solar entities; the third Logos controls the lunar entities and their correspondences elsewhere in the system.

This rule must not be carried too far in detail as long as man's mind is of its present calibre. The mystery lies in the realisation that all is carried on in a divine co-operation that has its base outside the system. Hence too the fact that the first Logos is called the Destroyer, because He is abstraction, if viewed from below upwards. His work is the synthesis of Spirit with Spirit, their eventual abstraction from matter, and their unification with their cosmic source. Hence also He is the one who brings about pralaya or the disintegration of form,—the form from which the Spirit has been abstracted.

If we carry the analogy down to the microcosm a glimpse can be gained of the same idea and hence ability to comprehend with greater facility. The Ego (being to the man on the physical plane what the Logos is to His system) is likewise the animating will, the destroyer of forms, the producer of pralaya and the One Who withdraws the inner spiritual man from out of his threefold body; he draws them to himself the centre of his little system. The Ego is extra-cosmic as far as the human being on the physical plane is concerned, and in the realisation of this fact may come elucidation of the true cosmic problem involving the Logos and "the spirits in prison," as the Christian puts it.

c. *His mode of action* is a driving forward; the will that lies back of evolutionary development is His, and He it is who drives Spirit onward through matter till it eventually emerges from matter, having achieved two things:

First, Added quality to quality, and therefore emerging plus the gained faculty that experience has engendered.

Second, Increased the vibration of matter itself by means of its own energy, so that matter at the moment of pralaya and obscuration will have two main characteristics,—activity, the result of the Law of Economy, and a dual magnetism which will be the result of the Law of Attraction.

All of these three concepts are governed by the Law of Synthesis, which is the law of a coherent will-to-be, persisting not only in time and space, but within a still vaster cycle.

These preliminary statements have been laid down in an endeavour to show the synthesis of the whole. In the use of words comes limitation, and a clouding of the idea; words literally veil or hide thoughts, detract from their clarity, and confuse them by expression. The work of the second and third Logoi (being the production of the objectivity of the essential Spirit) is more easy to grasp in broad outline than the more esoteric work of the first Logos, which is that of the animating will.   *A Treatise on Cosmic Fire*, pp. 148/50

• When the primordial ray of intelligent activity, the divine ray of intelligent love, and the third cosmic ray of intelligent will meet, blend, merge, and blaze forth, the Logos will take His fifth initiation, thus completing one of His cycles. When the rotary, the forward, and the spiral cyclic movements are working in perfect synthesis then the desired vibration will have been reached. When the three Laws of Economy, of Attraction, and of Synthesis work with perfect adjustment to each other, then nature will perfectly display the needed functioning, and the correct adaptation of the material form to the indwelling spirit, of matter to life, and of consciousness to its vehicle.
A Treatise on Cosmic Fire, p. 45
See also: 'Effects of Synthetic Motion',
A Treatise on Cosmic Fire, pp. 1128/52

11. ENERGY AND FORCE

• All that we surely know is that all forms are aspects of energy;

PHYSICAL FORCES OF EVOLUTION 241

that there is an interplay and an impact of energies upon our planet; that the planet is itself an energy unit composed of a multitude of energy units, and that man himself is likewise a composite bundle of forces and moving in a world of force. This is where science so wonderfully has led us, and this is where the astrologer, the occultist, the idealist and the mystic also meet and testify to a concealed Deity, to a living Being, to a Universal Mind, and to a central Energy.

The Labours of Hercules, pp. 11/2

• All forms are related, inter-related and interdependent; the planetary etheric body holds them together so that a cohesive, coherent, expressive Whole is presented to the eye of man, or one great unfolding consciousness to the perception of the Hierarchy. Lines of light pass from form to form. Some are bright and some are dim; some move or circulate with rapidity, others are lethargic and slow in their interplay; some seem to circulate with facility in some particular kingdom in nature and some in another; some come from one direction and some from a different one, but all are in movement all the time; it is a constant circulation. All are passing on and into and through, and there is not one single atom in the body which is not the recipient of this living, moving energy; there is no single form that is not "kept in shape and livingness" by this determined inflow and outflow, and there is therefore no part of the body of manifestation (which is an integral part of the planetary vehicle of the Lord of the World) which is not in complex but complete touch with HIS divine intention—through the medium of HIS three major centres: Shamballa, the Hierarchy, and Humanity.

Telepathy and the Etheric Vehicle, pp. 148/9

• The keynotes upon which the occult philosophy is built are:

1. There is naught in manifestation except organised energy
2. Energy follows or conforms itself to thought
3. The occultist works in energy and with energies.

The thought of God brought the universe of energies into organised form upon the highest of the seven planes, or upon the first cosmic etheric level. These energies have for untold aeons been directed from the fourth or lowest of the cosmic etheric planes, the plane which we call the buddhic and regard as the first definitely spiritual plane, in our usually erroneous thought; this direction has been under impression from Shamballa, and the Masters have "manipulated these energies in conformity with the Plan, which is the blueprint of the Purpose."   *The Externalisation of the Hierarchy*, p, 674

• Every atom is a focal point of force, the force of substance itself, the life or vitality of the third aspect, the life of that cosmic Entity Who is *to the Logos* the negative aspect of electricity.

Every form and aggregate of atoms, is simply a force centre produced by the action of positive force and its interaction with negative energy. It is the vitality of the second aspect working in conjunction with the third, and producing—in time and space—that illusion or maya which temporarily blazes forth, and attracts attention, creating the impression that matter is a concrete something. There is no such thing as concretion in reality; there is only force of different kinds, and the effect produced on consciousness by their interplay.

Back of all forms and of all substances (as yet but little contacted and realised) lies a third type of force, which utilises these two other factors to produce eventual harmony, and which is itself on its own plane the sumtotal of the second. It can be called:

a. The one synthesising Life.

b. Electric fire.

c. The point of equilibrium.

d. Unity or harmony.

e. Pure Spirit.

f. Dynamic Will.

g. Existence.

It is a Force, working through a dual manifestation of differentiated force, through the energy of matter, the coherency of forms, through force centres, and force points. It is FOHAT in triple demonstration, of which the final or third is as yet unknown and inconceivable. *A Treatise on Cosmic Fire*, pp. 527/9

• The Science of the Antahkarana is connected with the entire problem of energy, but peculiarly with the energy handled by the individual and with the forces by which the individual relates himself to other individuals or to groups. For the sake of clarity, we will give the name of

a. ENERGY: to all forces pouring into the individual form from whatever direction and source. To these major energies, the names of "sutratma" or "life thread" or "silver cord" have frequently been given.

b. FORCE: to all the energies which—after due manipulation and concentration—are projected by the individual or group in any direction and with many possible motives, some good and many selfish.

Education in the New Age, p. 143

• The Science of the Antahkarana deals, therefore, with the entire incoming system of energy, with the processes of usage and transformation and fusion. It deals also with the outgoing energies and their relationship to the environment and is the basis of the science of the force centres. The incoming and the outgoing energies constitute finally two great stations of energy, one characterised by power and the other by love, and all directed to the illumination of the individual and of humanity as a whole, through the medium of the Hierarchy composed of individuals. This is basically the Science of the Path.

Education in the New Age, pp. 147/8

• Certain wide generalisations anent the etheric body should be recalled at this point. The existence of an etheric body in relation to all tangible and exoteric forms is accepted today by many scientific schools; nevertheless the original teaching has been amended in order to bring it into line with the usual theories of energy and its forms of expression. Recognition is given today, by thinkers, to the factual nature of energy (and I am using that word "factual" most advisedly); energy is now regarded as all that IS; manifestation is the manifestation of a sea of energies, some of which are built into forms, others constitute the medium in which those forms live and move and have their being, and still others are in process of animating both the forms and their environing substantial media. It must also be remembered that forms exist within forms; this is the basis of the symbolism which is to be found in the intricate carved ivory balls of the Chinese craftsmen where ball within ball is to be discovered, all elaborately carved and all free and yet confined. You—as you sit in your room—are a form within a form; that room is itself a form within a house, and that house (another form) is probably one of many similar houses, placed the one on top of another or else side by side, and together composing a still larger form. Yet all these diverse forms are composed of tangible substance which—when coordinated and brought together by some recognised design or idea in the mind of some thinker—creates a material form. This tangible substance is composed of living energies, vibrating in relation to each other, yet owning their own quality and their own qualified life.

Telepathy and the Etheric Vehicle, pp. 177/8

• It is when the aspirant recognises that he himself is composed of energy units—held in coherent expression by a still stronger energy, that of integration—that he begins consciously to work in a world of forces similarly composed; he then begins to use energy of a certain kind, and selectively, and takes one of the initial steps towards becoming a true occultist. This world of energy in which he lives and moves and has his being is the living, organised vehicle of manifestation of the planetary Logos. Through it energies are circulating all the time and are in constant movement, being directed and controlled by the head centre of the planetary Logos; they create great vortices of force or major points of tension throughout His body of manifestation. The Spiritual Hierarchy of our planet is such a vortex; Humanity itself is another, and one which is today in a condition of almost violent activity, owing to its becoming a focus of divine attention.

The Rays and the Initiations, pp. 549/50

• Medical science, through its study of the nervous system and its recognition of the power of thought over the physical body, is moving rapidly in a right direction. When it admits, in relation to the physical body, that "energy follows thought," and then begins to experiment with the concept of thought currents (as they are erroneously called) which are directed to certain areas of the etheric body—where the esotericists posit the existence of energy points or centres—much will then be discovered.   *Esoteric Healing*, p. 282

• . . . it is important that all occultists should learn to interpret and to think in terms of energy and of force, in contradistinction to the sheaths or instruments employed. The mystic has recognised this "force" factor, but has only worked with the *positive* force aspect. The occultist must recognise and work with three types of force, or energy, and therein lies the distinction between his work and that of the mystic. He recognises:

1. Positive force....................Or that which energises.
2. Negative force..................Or that which is the recipient of energy; that which acts or assumes form under the impact of positive force.
3. Light, or harmonic force.....That which is produced by the union of these two. The result is *radiant energy,* and is the result of the equilibrising of the two others.

These three aspects of energy have been called, as has been often said:

a. Electric fire..........positive energy..........Father.
b. Fire by friction…..negative energy........Mother.
c. Solar fire..............radiant energy..........Sun or Son.

Each of these two last aspects demonstrates within itself in a dual manner, but the effect is a unified whole as regards the great Unity in which they are demonstrating.   *A Treatise on Cosmic Fire,* pp. 832/3

• [The aspirant]: He must study the laws of transmutation and be a student of that divine alchemy which will result in a knowledge of how to transmute the lower force into the higher, of how to transfer his consciousness into the higher vehicles, and of how to manipulate energy currents so that his own nature is transformed. He will then become a channel for the light of the Ego, and for the illumination of buddhi to pour through for the saving of the race, and the lighting of those who stumble in dark places. He must demonstrate the laws of radioactivity in his own life on the physical plane. His life must begin to radiate, and to have a magnetic effect upon others. By this I mean he will begin to influence that which is imprisoned in others, for he will reach—through his own powerful vibrations—the hidden centre in each one. I do not mean by this the physical or magnetic effect that many quite unevolved souls have upon others. I refer to that spiritual radiation that is only responded to and realised by those who themselves are becoming aware of the spiritual centre within the heart. At this stage the man is recognised as one who can speak occultly "heart to heart." He becomes a stimulator of the heart centre in his brother, and one who arouses men into activity for others.

A Treatise on Cosmic Fire, p. 863

12. ATTRACTION–REPULSION–COHESION

• [The Law of Attraction]: This law is, as we know, the basic law of all manifestation, and the paramount law for this solar system. It might strictly be called the Law of Adjustment or of Balance, for it conditions that aspect of electrical phenomena which we call *neutral*. The Law of Economy is the basic law of one pole, that of the negative aspect; the Law of Synthesis is the basic law of the positive pole, but the Law of Attraction is the law for the fire which is produced by the merging during evolution of the two poles. From the standpoint of the human being, it is that which brings about the realisation of self-consciousness; from the point of view of the subhuman beings it is that which draws all forms of life on to self-realisation; whilst in connection with the superhuman aspect it may be stated that this law of life expands into the processes conditioned by the higher law of Synthesis, of which the Law of Attraction is but a subsidiary branch.

Strictly speaking, the Law of Attraction is a generic term under which are grouped several other laws similar in nature but diverse in their manifestations. It might be useful if we enumerated a few of these laws, thereby enabling the student to get (as he studies them in their totality) a broad general idea as to the Law and its modifications, its spheres of influence and the scope of its activity. It should be noted here as a basic proposition in connection with all atoms that the Law of Attraction governs the Soul aspect. *The Law of Economy is the law of the negative electron; the Law of Synthesis is the Law of the positive central life; whilst the Law of Attraction governs that which is produced by the relation of these two*, and is itself controlled by a greater cosmic law which is the principle of the intelligence of substance. It is the law of Akasha.

It must be borne in mind that these three laws are the expression of the intent or purpose of the three Logoic Aspects. The Law of Economy is the governing principle of Brahma or the Holy Spirit; the Law of Synthesis is the law of the Father's life; whilst the Son's life is governed by, and manifests forth divine attraction. Yet these three are the three subsidiary laws of a greater impulse which governs the life of the Unmanifested Logos. . . .

The subsidiary aspects, or laws, of the Law of Attraction might be enumerated as follows:

1. *The Law of Chemical Affinity.* . . .
2. *The Law of Progress.* . . .
3. *The Law of Sex.* . . .
4. *The Law of Magnetism.* . . .
5. *The Law of Radiation.* . . .
6. *The Law of the Lotus.* . . .
8. *The Law of Gravitation.* . . .

9. *The Law of Planetary Affinity.* . . .
10. *The Law of Solar Union.* . . .
11. *The Law of the Schools.* . . .

A Treatise on Cosmic Fire, pp. 1166/74
See also: 'The Law of Attraction – The Subsidiary Laws',
A Treatise on Cosmic Fire, pp. 1166/85

• *The Law of Cohesion.*—This is one of the branch laws of the cosmic Law of Attraction. It is interesting to notice how this law demonstrates in this Love-System in a threefold manner:

On the plane of the Monad, as the law of cohesion, the law of birth, if we might use that term, resulting in the appearance of the Monads in their seven groups. Love the source, and the Monad of love, the result.

On the plane of buddhi, as the law of magnetic control. It shows itself as the love-wisdom aspect, irradiating the ego, and eventually gathering to itself the essence of all experience, garnered, via the Ego, through the personality lives, and controlled throughout from the plane of buddhi. . . .

On the astral plane, as love demonstrating through the personality. All branches of the law of attraction, demonstrating in this system, show themselves as a force that ingathers, that tends to coherence, that results in adhesion, and leads to absorption. All these terms are needed to give a general idea of the basic quality of this law.

This law is one of the most important of the systemic laws, if it is permissible to differentiate at all; we might term it the law of coalescence.

On the path of involution it controls the primal gathering together of molecular matter, beneath the atomic subplane. It is the basis of the attractive quality that sets in motion the molecules and draws them into the needed aggregations. It is the measure of the subplanes. The atomic subplane sets the rate of vibration; the Law of Cohesion might be said to fix the colouring of each plane. It is the same thing in other words. We need always to remember in discussing these abstract fundamentals that words but dim the meaning, and serve but as suggestions and not as elucidations.

A Treatise on Cosmic Fire, pp. 576/7

• *Capacity to cohere*. This is the ability of all intelligent, active Lives during evolution to conform to the Law of Attraction and Repulsion, and thus form a conscious, intelligent part of a greater life. It is literally the transmutation of manas into wisdom. Though all that IS exists in form yet little is as yet brought under the intelligent control of the entity within the form. Only the Heavenly Men and Their superior embracing lives are consciously and intelligently working through and dominating the form, for only They, as yet, are perfected manas. Beneath them come many grades of consciousness. Man is gradually achieving that conscious control over matter in the three worlds which his divine Prototypes, the Heavenly Men, have already achieved. They are attaining a similar control on higher levels. Below man come many lives who are blind and unconscious of the congery or subdivision of which they form part. Thus can be seen, in general outline, the place of manas at the present stage.

A Treatise on Cosmic Fire, pp. 409/10

• Cohesion is also plainly to be seen as the distinguishing feature of our present system, the second. It is the aim of all things to unite; approximation, unification, a simultaneous attraction between two or more is ever to be seen as a governing principle, whether we look at the sex problem, or whether it demonstrates in business organisation, in scientific development, in manufacture, or in politics. Well might we say that the *At-one-ment* of the many separated is the keynote of our system. *A Treatise on Cosmic Fire*, p. 579

• *Love* is the right apprehension of the uses and purposes of form, and of the energies involved in form-building, the utilisation of form, and the eventual dissipation of the superseded form. It involves a realisation of the Laws of Attraction and Repulsion, of the magnetic interplay between all forms, great and small, of group relationships, of the galvanizing power of the unifying life, and the attractive power of one unit upon another, be it atom, man, or solar system. It involves an understanding of all forms, form purposes, and form relationships; it concerns the building processes in man himself, and in the solar system; and it necessitates the development of those powers within man which will make him a conscious Builder, a solar Pitri of a coming cycle. This is one of the great revelations at initiation: the unveiling to the initiate of the particular cosmic centre whence emanates the type of force or energy which he, the initiate, will be concerned with when he becomes in due course of time a solar Pitri, or divine manasaputra to coming humanity. Hence he must have, not only knowledge, but the energy of love likewise to enable him to perform the function of linking the higher three and the lower four of a future race of men at some distant period, thus permitting of their individualising through the *sacrifice of his own fully conscious middle principle*.

Treatise on Cosmic Fire, pp. 881/2

• The activity of the second Logos is carried on under the cosmic Law of Attraction. The Law of Economy has for one of its branches a subsidiary Law of marked development called the Law of Repulsion. The cosmic Laws of Attraction and Economy are therefore the *raison d'être* (viewed from one angle) of the eternal repulsion that goes on as Spirit seeks ever to liberate itself from form. The matter aspect always follows the line of least resistance, and repulses all tendency to group formation, while Spirit, governed by the Law of Attraction, seeks ever to separate itself from matter by the method of attracting an ever more adequate type of matter in the process of distinguishing the real from the unreal, and passing from one illusion to another until the resources of matter are fully utilised.

Eventually the Indweller of the form feels the urge, or attractive pull, of its Own Self. The reincarnating jiva, for instance, lost in the maze of illusion, begins in course of time to recognise (under the Law of Attraction) the vibration of its own Ego, which stands to it as the Logos of its own system, its deity in the three worlds of experience. Later, when the body egoic itself is seen as illusion, the vibration of the Monad is felt, and the jiva, working under the same law, works its way back through the matter of the two planes of superhuman evolution, till it is merged in its own essence.

A Treatise on Cosmic Fire, pp. 144/5

- We have referred to the Law of Repulsion as one of the subsidiary branches of the great Law of Economy, which governs matter. Repulsion is brought about by rotary action, and is the basis of that separation which prevents the contact of any atom with any other atom, which keeps the planets at fixed points in space and separated stably from each other; which keeps them at a certain distance from their systemic centre, and which likewise keeps the planes and subplanes from losing their material identity. Here we can see the beginning of that age-long duel between Spirit and matter, which is characteristic of manifestation, one aspect working under the Law of Attraction, and the other governed by the Law of Repulsion. From aeon to aeon the conflict goes on, with matter becoming less potent. Gradually (so gradually as to seem negated when viewed from the physical plane) the attractive power of Spirit is weakening the resistance of matter till, at the close of the greater solar cycles, destruction (as it is called) will ensue, and the Law of Repulsion be overcome by the Law of Attraction. It is a destruction of form and not of matter itself, for matter is indestructible. This can be seen even now in the microcosmic life, and is the cause of the disintegration of form, which holds itself as a separated unit by the very method of repulsing all other forms. It can be seen working out gradually and inappreciably in connection with the *Moon*, which no longer is repulsive to the earth, and is giving of her very substance to this planet. H. P. B. hints at this in *The Secret Doctrine*, and I have here suggested the law under which this is so. *A Treatise on Cosmic Fire*, p. 154

- *Law 4. The Law of Repulsion.* This law concerns itself with the ability of an atom to throw off, or refuse to contact, any energy deemed inimical to group activity. It is literally a law of service, but only comes consciously into play when the atom has established certain basic discriminations, and guides its activities through a knowledge of the laws of its own being. This law is not the same as the Law of Repulsion which is used in connection with the Law of Attraction between forms which have relation to the material. The laws we are now considering have relation to the psyche, or to the Vishnu aspect. One group of laws concern energies emanating from the physical sun; the ones we are now considering emanate from the heart of the Sun. The "repulsion" here dealt with has the effect (when consciously applied through the developed heart energy of a human atom, for instance) of furthering the interests of the repulsed unit and of driving this unit closer to its own centre. Perhaps some idea of the great beauty of this law as it works out can be gathered from an occult phrase in a certain old book:

"This repulsive force drives in seven directions, and forces all that it contacts back to the bosom of the seven spiritual fathers."

Through repulsion, the units are driven home and the straying unconscious ones are forced towards their own centre. The Law of Repulsion, or the stream of energy for which it is but a name, can work from any centre, but as dealt with here, it must *emanate from the heart* if it is to bring about the necessary group work.

A Treatise on Cosmic Fire, p. 1217

• *Repetition of Cyclic Action is Governed by Two Laws:*
Perhaps it is more accurate to say that it is governed by one law, primarily, and a subsidiary law. This leads to two general types of cycles, and is involved in the very nature of the Self and of the not-self. The interplay of the two by the aid of mind produces that which we call environment or circumstance.

The general law, which produces cyclic effect, is the Law of Attraction and Repulsion, of which the subsidiary law is the Law of Periodicity, and of Rebirth. Cyclic evolution is entirely the result of the activity of matter, and of the Will or Spirit. It is produced by the interaction of active matter and moulding Spirit. Every form holds hid a Life. Every life constantly reaches out after the similar life latent in

PHYSICAL FORCES OF EVOLUTION

other forms. When Spirit and matter sound the same note evolution will cease. When the note sounded by the form is stronger than that of Spirit, we have attraction between *forms*. When the note sounded by Spirit is stronger than that of matter and form, we have Spirit repelling form. Here we have the basis for the battlefield of life, and its myriads of intermediate stages, which might be expressed as follows:

- *a.* The period of the domination of the form note is that of involution.
- *b.* The period of the repulsion of form by Spirit is that of the battlefield of the three worlds.
- *c.* The period of the attraction of Spirit and Spirit, and the consequent withdrawal from form is that of the Path.
- *d.* The period of domination of the note of Spirit is that of the higher planes of evolution.

To the synchronisation of the notes, or to the lack of synchronisation, may be attributed all that occurs in the world cycles. Thus we have the production of harmony; first, the basic note of matter, then the note of Spirit gradually overcoming the lower note and usurping attention till gradually the note of Spirit overpowers all other notes. Yet it must be borne in mind that it is the note of the life that holds the form together. The note of the Sun, for instance, holds in just attraction the circling spheres, the planets. The notes synchronise and harmonise till the stage of adequacy is reached and the period of abstraction. Cyclic evolution proceeds. A human being, similarly, holds (by means of his note) the atoms of the three bodies together, being to them as the central sun to the planets. Primarily, nevertheless, it may be posited that the Law of Attraction is the demonstration of the powers of Spirit, whilst the Law of Repulsion governs the form. Spirit attracts Spirit throughout the greater cycle. In lesser cycles, Spirit temporarily attracts matter. The tendency of Spirit is to merge and blend with Spirit. Form repulses form, and thus brings about separation. But—during the great cycle of evolution—when the third factor of Mind comes in, and when the point of balance is the goal, the cyclic display of the interaction between Spirit and form is seen, and the result is the ordered cycles of the planets, of a human being, and of an atom. Thus, through repetition, is consciousness developed, and responsive faculty induced. When this faculty is of such a nature that it is an inherent part of the Entity's working capital, it has to be exercised on every plane, and again cyclic action is the law, and hence rebirth again and again is the method of exercise. When the innate conscious faculty of every unit of consciousness has become co-ordinated as part of the equipment of the Logos on every plane of the solar system, then, and only then, will cyclic evolution cease, will rotary movement on every plane of the cosmic physical plane be of such a uniform vibration as to set up action on the next cosmic plane, the astral. *A Treatise on Cosmic Fire,* pp. 274/6

13. EXPANSION

• We speak at times of an expanding universe; what we really mean is an expanding consciousness, for this etheric body of the Entity, Space, is the recipient of many types of informing and penetrating energies, and it is also the field for the intelligent activity of the indwelling Lives of the Universe, of the many constellations, of the distant stars, of our solar system, of the planets within the system, and of all that constitutes the sum total of these separated living forms. The factor which relates them is consciousness and nothing else, and the field of conscious awareness is created through the interplay of all living intelligent forms within the area of the etheric body of that great Life which we call SPACE.

Telepathy and the Etheric Vehicle, p. 179

• *The Law of Expansion.*

This law of a gradual evolutionary expansion of the consciousness indwelling every form is the cause of the spheroidal form of every life in the entire solar system. It is a fact in nature that all that is in existence dwells within a sphere. The chemical atom is spheroidal; man dwells within a sphere, as does the planetary Logos and the solar Logos, this sphere being the form matter takes when its own internal activity, and the activity of the form are working in unison. It requires the two types of force—rotary and spiral-cyclic—to produce this. Scientists are beginning to recognise this more or less, and to realise that it is the Law of Relativity, or the relation between all atoms, which produces that which is called Light, and which, in its aggregated phenomena, forms that composite sphere, a solar system. The motion of the constellations *external* to the solar sphere is responsible for its form in conjunction with its own rotary motion in space. As the wave lengths of the light from the constellations, and their relation to the sun are better understood, and as the effect of those wave lengths or light vibrations (which are either attractive to, or repulsive to, the sun) are understood, much will be revealed. Little has as yet been grasped as to the effect those constellations in the heavens (which are antagonistic to the solar system), have upon it, and whose wave lengths it will not transmit, whose rays of light do not pierce (if it might be expressed in so unscientific a manner) through the solar periphery.

We are told in the *Secret Doctrine* that "the seven solar Rays dilate to seven suns and set fire to the whole cosmos." This it is which produces that final burning which ushers in the great pralaya, and brings to an end the logoic incarnation. It is produced under this Law of Expansion, and causes that eventual merging and blending of the seven sacred planetary schemes which marks the achievement of the goal, and their eventual perfection.

In occult literature this term "Law of Expansion" is limited to the discussion of the seven Rays, and to the subject of the *planetary* initiations. When dealing with the expansions of consciousness of the human being, and his initiations, we group them under the second "Law of Monadic Return."

Students should here remember that we are dealing with the expansions of consciousness of a planetary Logos through the medium of:

a. The chains.

b. The rounds.

c. The kingdoms of nature.

d. The root races.

It should be remembered that the consciousness He is in process of developing is that of the absolute will and purpose of the solar Logos, as it is the expression of the *desire* of the cosmic Logos.

A Treatise on Cosmic Fire, pp. 1040/2

14. SPACE AND TIME

• The Ancient Wisdom teaches that "space is an entity." It is with the life of this entity and with the forces and energies, the impulses and the rhythms, the cycles and the times and seasons that esoteric astrology deals. . . .

Space is an entity and the entire "vault of heaven" (as it has been poetically called) is the phenomenal appearance of that entity. You will note that I did not say the material appearance, but the phenomenal appearance. Speculation about the nature, the history and identity of that entity is useless and of no value. Some dim idea, providing analogy even when eluding specifications, might be gained if you will endeavour to think of the human family, the fourth kingdom in nature, as an entity, as constituting a single unit, expressing itself through the many diversified forms of man. You, as an individual, are an integral part of humanity, yet you lead your own life, you react to your own impressions, you respond to exterior influences and impacts, and in your turn you emanate influences, send forth some form of character radiation and express some quality or qualities. You thereby, and in some measure, affect your environment and those whom you contact. Yet all the while you remain part of a phenomenal entity to which we give the name of *humanity*. Now extend this idea to a greater phenomenal entity, the solar system. This entity is itself an integral part of a still greater life which is expressing Itself through seven solar systems, of which ours is one. If you can grasp this idea, a vague picture of a great underlying esoteric truth will emerge into your consciousness. It is the life and the influence, the radiations and emanations of this entity, and their united effect on our planetary life, the kingdoms in nature and the unfolding human civilizations, which we shall have briefly to consider. . . .

The next point for each of you to grasp is the fact that the ether of space is the field in and through which the energies from the many originating Sources play. We are, therefore, concerned with the etheric body of the planet, of the solar system, and of the seven solar systems of which our system is one, as well as with the general and vaster etheric body of the universe in which we are located. I employ the word "located" here with deliberation and because of the inferences to which it leads. This vaster field, as well as the smaller and more localised fields, provides the medium of transmission for all the energies which play upon and through our solar system, our planetary spheres and all forms of life upon those spheres. It forms one unbroken field of activity in constant ceaseless motion—an eternal medium for the exchange and transmission of energies.

Esoteric Astrology, pp. 7/10

• This etheric body—vast and unknown as it is, as to its extent—is nevertheless limited in nature and static (relatively speaking) in capacity; it preserves a set form, a form of which we know absolutely nothing, but which is the etheric form of the Unknown Entity. To this form the esoteric science gives the name of SPACE; it is the fixed area in which every form, from a universe to an atom, finds its location.

Telepathy and the Etheric Vehicle, pp. 178/9

• It might be profitable to point out that the entire universe is etheric and vital in nature and of an extension beyond the grasp of the greatest mind of the age, mounting into more than astronomical figures—if that statement even conveys sense to your minds. This extent cannot be computed, even in terms of light years; this cosmic etheric area is the field of untold energies and the basis of all astrological computations; it is the playground of all historical cycles—cosmic, systemic and planetary—and is related to the constellations, to the worlds of suns, to the most distant stars and to the numerous recognised universes, as well as to our own solar system, to the many planets, and to that planet upon which and in which we move and live and have our being, as well as to the smallest form of life known to science and perhaps covered by the meaningless term "an atom." All are found existing in Space—Space is etheric in nature and—so we are told in the occult science—Space is an Entity. The glory of man lies in the fact that he is aware of space and can imagine this space as the field of divine living activity, full of active intelligent forms, each placed in the etheric body of this unknown Entity, each related to each other through the potency which not only holds them in being but which preserves their position in relation to each other; yet each of these differentiated forms possesses its own differentiated life, its own unique quality or integral colouring, and its own specific and peculiar form of consciousness. *Telepathy and the Etheric Vehicle*, p. 178

• (Sutra 54): This intuitive knowledge, which is the great Deliverer, is omnipresent and omniscient and includes the past, the present and the future in the Eternal Now.

The only part of this sutra which is not clear even to the superficial reader is the significance of the words Eternal Now, and these it is not possible to comprehend until soul-consciousness is developed. To say that time is a succession of states of consciousness and that the present is lost in the past instantaneously, and merged in the future as it is experienced, is of small avail to the average student. To say that there is a time when sight is lost in vision, when the sum total of life anticipations are realized in a moment of accomplishment and that this persists forever, and to point to a state of consciousness in which there is no sequence of events and no succession of realizations is to speak in a language of mystery. Yet so it is and will be. When the aspirant has reached his goal he knows the true significance of his immortality and the true nature of his liberation. Space and time become for him meaningless terms. The only true Reality is seen to be the great central life force, remaining unchanged and unmoved at the centre of the changing evanescent temporal forms.

"I am," says the human unit and regards himself as the self, and identifies himself with the changing form. Time and space are for him the true realities. "I am That," says the aspirant and seeks to know himself as he truly is, a living word, part of a cosmic phrase. For him space no longer exists; he knows himself as omnipresent. "I am That I am," says the freed soul, the liberated man, the Christ. Neither time nor space exist for him, and omniscience and omnipresence are his distinctive qualities. *The Light of the Soul,* pp. 366/7

• As we well know, the nature of the motion on the plane of matter is *rotary*. Each atom of matter rotates on its own axis, and each larger atom, from the purely physical standpoint, likewise does the same; a cosmic atom, a solar system, a planetary atom, and a human atom, man, can be seen equally rotating at differing degrees of velocity upon their own axis or around their own pole. When we arrive at the plane of mind, and have to consider the activity of the second aspect of divinity, that which builds and holds the forms in coherent form, and which is the basis of the phenomenon we call *time* (literally, the awareness of the form), a different type of force or motion becomes apparent. This type of energy in no way negates or renders useless the atomic rotary type, but involves it, and yet at the same time it brings the atoms of all degrees under the influence of its own activity, so that in every form which is in manifestation, the two types are manifested. . . .

The activity of the second aspect has been called *spiral-cyclic*, which in itself involves the concept of duality. This activity is the cause of all cyclic evolution, and has been called in the occult phraseology "the activity of Brahma's year." It is that which brings about the periodical appearing and disappearing of all existences, great or small. It is intimately linked with the will aspect of Divinity, and with the Lipika Lords of the highest degree and its origin is, therefore, difficult for us to comprehend. Perhaps all that can be said about it is that it is largely due to certain impulses which (as far as our solar system is concerned) can be traced to the sun Sirius. These impulses find their analogy in the impulses emanating in cyclic fashion from the causal body of man, which impulses bring about his appearance upon the plane of maya for a temporary period.

A Treatise on Cosmic Fire, pp. 1032/3

• You might here ask and rightly so: What is this plan? When I speak of the plan I do not mean such a general one as the plan of evolution or the plan for humanity which we call by the somewhat unmeaning term of soul unfoldment. These two aspects of the scheme for our planet are taken for granted, and are but modes, processes and means to a specific end. The plan as at present sensed, and for which the Masters are steadily working, might be defined as follows:—It is the production of a subjective synthesis in humanity and of a telepathic interplay which will eventually annihilate time. It will make available to every man all past achievements and knowledges, it will reveal to man the true significance of his mind and brain and make him the master of that equipment and will make him therefore omnipresent and eventually open the door to omniscience. This next development of the plan will produce in man an understanding —intelligent and cooperative—of the divine purpose for which the One in Whom we live and move and have our being has deemed it wise to submit to incarnation.

A Treatise on White Magic, pp. 403/4

• Attraction of matter to Spirit and the building of a form for the use of Spirit is the result of electrical energy in the universe, which in each case brings the lesser lives or spheres into its range of influence. The magnetic force, the life of the Logos gathers together His body of manifestation. The magnetic force of the Heavenly Man, the planetary Logos, gathers out of the solar ring-pass-not that which He needs for each incarnation.

The magnetic force of the Ego gathers, at each rebirth, matter within the particular sphere or scheme within which the Ego has place. So on down the scale, we find the lesser pursuing its round ever within the greater.

Therefore we have (during a period of Attraction and Repulsion, or a life cycle) that which we call Time and Space, and this holds equally true in the life cycle of a Logos or an ant, or a crystal. There are cycles of activity in matter, due to some energising Will, and then Time and

Space are known. There are cycles of non-being when Time and Space are not, and the energising Will is withdrawn. But we must not forget that this is purely relative, and only to be considered from the standpoint of the particular life or entity involved, and the special stage of awareness reached. All must be interpreted in terms of consciousness.

A Treatise on Cosmic Fire, pp. 283/4

• It might be of value here if I clarified my use of the words "higher ego." As you know, if you have read Esoteric Psychology Vols. I and II, the soul is an aspect of the divine energy in time and space. We are told that the Solar Logos circumscribed for His use and for the meeting of His desire, a certain measure of the substance of space and informed it with His life and consciousness. He did this for His good purposes and in conformity with His self-realised plan and intent. Thus He submitted Himself to limitation. The human monad followed the same procedure and—in time and space—limited itself in a similar manner. On the physical plane and in the physical body, this phenomenal and transient entity controls its phenomenal appearance through the two aspects of life and consciousness.

Education in the New Age, p. 18

• It will be realised before long that time is entirely a brain event; a study of the sense of speed as registered by the brain, plus the capacity or incapacity of a human being to express this speed, will, when properly approached, reveal much that today remains a mystery.

The Destiny of the Nations, p. 32

• Time to the occultist is that cycle, greater or lesser, in which some life runs some specific course, in which some particular period begins, continues, and ends, in connection with the awareness of some Entity, and is recognised only as time when the participating life has reached a considerable stage of awareness. Time has been defined as a succession of states of consciousness, and it therefore may be studied from the point of view of :

a. Logoic consciousness, or the successive states of divine realisation within the solar sphere.

b. Planetary consciousness, or the consciousness of a Heavenly Man as He cycles successively through the scheme.

c. Causal consciousness, or the successive expanding of the intelligent awareness of a human being from life to life.

d. Human consciousness, or the awareness of a man on the physical plane, and progressively on the emotional and the mental planes.

e. Animal, vegetable and mineral consciousness which differs from the human consciousness in many particulars, and primarily in that it does not co-ordinate, or deduce and recognise separate identity. It resembles human consciousness in that it covers the response to successive contacts of the units involved during their small cycles.

f. Atomic consciousness, demonstrating through successive states of repulsion and attraction. In this last definition lies the key to the other states of consciousness.

A Treatise on Cosmic Fire, pp. 278/9

- WHY IS THE PROGRESS OF EVOLUTION CYCLIC

This question is one which necessarily appals us and makes us wonder.

Let us, therefore, deal with it as follows: Certain ideas are involved in the thought of cyclic progression, and these ideas it might pay us well to contemplate.

1. *The Idea of Repetition.*

This repetition involves the following factors:

a. Repetition in time: The thought of cyclic activity necessitates periods of time of differing length—greater or lesser cycles—but (according to their length) of uniform degree. A manvantara, or Day of Brahma, is always of a certain length, and so is a mahamanvantara. The cycles wherein an atom of any plane revolves upon its axis are uniform on its own plane.

b. Repetition in fact: This involves the idea of a key measure, or sound of any particular group of atoms that go to the composition of any particular form. This grouping of atoms will tend to the make-up of a particular series of circumstances and will repeat the measure or sound when an animating factor is brought to bear upon them. When the vitalising force is contacting at stated periods a certain set of atoms, it will call forth from them a specific sound which will demonstrate objectively as environing circumstances. In other words, the interplay of the Self and the not-self is invariably of a cyclic nature. The same quality in tone will be called forth by the Self as it indwells the form, but the key will ascend by gradual degrees. It is similar to the effect produced in striking the same note in different octaves, beginning at the base.

c. Repetition in space: This concept is involved deep in the greater concept of karma, which is really the law that governs the matter of the solar system, and which commenced its work in earlier solar systems. We have, therefore, *cycles in order,* and repetition in an ever-ascending spiral, under definite law.

The thoughts thus conveyed might be expressed likewise as follows:

a. The solar system repeating its activity......Repetition in Space.
b. A planetary chain repeating its activity....Repetition in Time.
c. The constant consecutive reverberation
  of a plane note, of a subplane note,
  and of all that is called into objectivity
  by that note..Plane Repetition.
d. The tendency of atoms to perpetuate
  their activity, and thus produce similarity
  of circumstance, of environment and of
  vehicle..Form Repetition.

When we carry these ideas on to every plane in the solar system, and from thence to the cosmic planes, we have opened up for ourselves infinitude. *A Treatise on Cosmic Fire*, pp. 273/4

• There is an aspect of human consciousness which has for long baffled the materialistic psychologist, and this is the curious power of prevision, the ability to foresee and foretell with accuracy events coming in the immediate future, or distant happenings. There are warnings given by some inner monitor which have again and again saved man from death and disaster; there are the appearances, to their friends and relatives, of men or women who have just died, before any word of their death has been received. This is not in the field of telepathic knowledge of the death, but involves the appearance of the person. There is the power to participate in events in distant places and to recover the recollection of what transpired with accuracy as to place, personnel and detail. These powers and many similar previsions and recognitions have long bewildered investigators and must find correct explanation. In their wise investigation, in the accumulation of responsible evidence, and in the later substantiation of the prevision, it will begin to be seen that some factor exists in man which is not bound by space-time limitations, but which transcends the normal human consciousness. The present attempted investigation and explanations are inadequate and do not account satisfactorily for all the facts. When, however, they are approached from the standpoint of the soul, with its faculty of omniscience and its freedom from categories of past, present and future (for they are lost in the consciousness of the Eternal Now), we shall begin to understand the process a little more clearly. When the true Dweller in the body is recognised and the laws of prevision are discovered, and when the power to foresee is generally prevalent, then we shall begin to find ample proof of the existence of the soul. It will be impossible to account for the ordinary phenomena then current without admitting its existence.

Esoteric Psychology I, pp. 103/4
See also: 'The Nature of Space',
Telepathy and the Etheric Vehicle, pp. 177-181

XIV. RADIATION AND TRANSMUTATION

• *Manas in the Final Rounds.*

   a. The transmutative process. Transmutation is a subject that from the earliest ages has occupied the attention of students, scientists and alchemists. The power to change, through the application of heat, is of course universally recognised, but the key to the mystery, or the secret of the systemic formula is advisedly guarded from all searchers, and is only gradually revealed after the second Initiation. The subject is so tremendous that it is only possible to indicate in broad general outlines how it may be approached. The mind of the public turns naturally to the transmutation of metals into gold with the aim in view of the alleviation of poverty. The mind of the scientist seeks the universal solvent which will reduce matter to its primordial substance, release energy, and thus reveal the processes of evolution, and enable the seeker to build for himself (from the primordial base) the desired forms. The mind of the alchemist searches for the Philosopher's Stone, that effective transmuting agent which will bring about revelation, and the power to impose the will of the chemist upon the elemental forces, which work in, by, and through matter. The religious man, especially the Christian, recognises the psychic quality of this transmutative power, and frequently speaks in the sacred books, of the soul being tried or tested seven times in the fire. All these students and investigators are recognising one great truth from their own constricted angle, and the whole lies not with one or another, but in the aggregate.

   In defining transmutation as it is occultly understood, we might express it thus: *Transmutation is the passage across from one state of being to another through the agency of fire.* The due comprehension of this is based on certain postulates, mainly four in number. These postulates must be expressed in terms of the Old Commentary, which is so worded that it reveals to those who have eyes to see, but remains enigmatic to those who are not ready, or who would misuse the knowledge gained for selfish ends. The phrases are as follows:

I. He who transfers the Father's life to the lower three seeketh the agency of fire, hid in the heart of Mother. He worketh with the Agnichaitans, that hide, that burn, and thus produce the needed moisture.

II. He who transfers the life from out the lower three into the ready fourth seeketh the agency of fire hid in the heart of Brahma. He worketh with the forces of the Agnishvattas, that emanate, that blend, and thus produce the needed warmth.

III. He who transfers the life into the gathering fifth seeketh the agency of fire hid in the heart of Vishnu. He worketh with the forces of the Agnisuryans, that blaze, that liberate the essence, and thus produce the needed radiance.

IV. First moisture, slow and all enveloping; then heat with ever-growing warmth and fierce intensity; then force that presses, drives and concentrates. Thus is radiance produced; thus the exudation; thus mutation; thus change of form. Finally liberation, escape of the volatile essence, and the gathering of the residue back to primordial stuff.

He who ponders these formulas and who meditates upon the method and suggested process will receive a general idea of the evolutionary process of transmutation which will be of more value to him than the formulas whereby the devas transmute the various minerals.

Transmutation concerns the life of the atom, and is hidden in a knowledge of the laws governing radioactivity. It is interesting to note how in the scientific expression 'radioactivity,' we have the eastern conception of Vishnu-Brahma, or the Rays of Light vibrating through matter. Hence the usually accepted interpretation of the term 'atom' must be extended from that of the atom of chemistry to include:

a. All atoms or spheres upon the physical plane.

b. All atoms or spheres upon the astral and mental planes.

c. The human being in physical incarnation.

d. The causal body of man on its own plane.

e. All planes as entified spheres.

f. All planets, chains and globes within the solar system.

g. All monads on their own plane, whether human monads or Heavenly Men.
   h. The solar Ring-Pass-Not, the aggregate of all lesser atoms.

In all these atoms, stupendous or minute, microcosmic or macrocosmic, the central life corresponds to the positive charge of electrical force predicated by science, whether it is the life of a cosmic Entity such as a solar Logos, or the tiny elemental life within a physical atom. The lesser atoms which revolve round their positive centre, and which are at present termed electrons by science, are the negative aspect, and this is true not only of the atom on the physical plane, but of the human atoms, held to their central attractive point, a Heavenly Man, or the atomic forms which in their aggregate form the recognised solar system. All forms are built up in an analogous manner and the only difference consists—as the text-books teach—in the arrangement and the number of the electrons. The electron itself will eventually be found to be an elemental, tiny life.

The second point I seek to make now is: *Radiation is transmutation in process of accomplishment.* Transmutation being the liberation of the essence in order that it may seek a new centre, the process may be recognised as radioactivity technically understood and applied to all atomic bodies without exception.

That science has but recently become aware of radium (an example of the process of transmutation) is but the fault of science. As this is more comprehended it will be found that all radiations, such as magnetism or psychic exhalation, are but the transmuting process proceeding on a large scale. The point to be grasped here is that the transmuting process, when effective, is superficially the result of outside factors. Basically it is the result of the inner positive nucleus of force or life reaching such a terrific rate of vibration, that it eventually scatters the electrons or negative points which compose its sphere of influence, and scatters them to such a distance that the Law of Repulsion dominates. They are then no more attracted to their original centre but seek another. The atomic sphere, if I might so express it, dissipates, the electrons come under the Law of Repulsion, and the central essence escapes and seeks a new sphere, occultly understood.

We must remember always that all within the solar system is dual, and is in itself both negative and positive: positive as regards its own form, but negative as regards its greater sphere. Every atom therefore is both positive and negative,—it is an electron as well as an atom.

Therefore, the process of transmutation is dual and necessitates a preliminary stage of application of external factors, a fanning and care and development of the inner positive nucleus, a period of incubation or of the systematic feeding of the inner flame, and an increase of voltage. There is next a secondary stage wherein the external factors do not count so much, and wherein the inner centre of energy in the atom may be left to do its own work. These factors may be applied equally to all atoms; to the mineral atoms which have occupied the attention of alchemists so much, to the atom, called man who pursues the same general procedure being governed by the same laws; and to all greater atoms, such as a Heavenly Man or a solar Logos.

The process might be tabulated as follows:

1. The life takes primitive form.
2. The form is subjected to outer heat.
3. Heat, playing on the form, produces exudation and the factor of moisture supervenes.
4. Moisture and heat perform their function in unison.
5. Elemental lives tend all lesser lives.
6. The devas co-operate under rule, order and sound.
7. The internal heat of the atom increases.
8. The heat of the atom mounts rapidly and surpasses the external heat of its environing.
9. The atom radiates.
10. The spheroidal wall of the atom is eventually broken down.
11. The electrons or negative units seek a new centre.
12. The central life escapes to merge with its polar opposite becoming itself negative and seeking the positive.
13. This is occultly obscuration, the going-out of the light tem-

porarily, until it again emerges and blazes forth.

More detailed elucidation will not be possible here nor advisable:

It will be apparent, therefore, that it should be possible, from the standpoint of each kingdom of nature, to aid the transmuting process of all lesser atoms. This is so, even though it is not recognised; it is only when the human kingdom is reached that it is possible for an entity consciously and intelligently to do two things:

First: aid in the transmutation of his own positive atomic centre from the human into the spiritual.

Second: assist at the transmutation

a. From the lower mineral forms into the higher forms.
b. From the mineral forms into the vegetable.
c. From vegetable forms into the animal forms.
d. From animal forms into the human or consciously and definitely to bring about individualisation.

That it is not done as yet is due to the danger of imparting the necessary knowledge. . . .

Man will eventually work with the three kingdoms but, only when brotherhood is a practice and not a concept.

A Treatise on Cosmic Fire, pp. 475/80
See also: 'Manas in the Final Rounds',
A Treatise on Cosmic Fire, pp. 475/504

• *The Cause of Radiation.* The student will only be able to get a true view of this matter if he views the subject in a large way. Two aspects of the matter naturally come before his mental vision, both of which must be dealt with if any adequate concept of this subject is to be reached,—a subject which has engrossed philosophers, scientists and alchemists for hundreds of years consciously or unconsciously. We must, therefore, consider:

a. That which radiates.
b. That which is the subjective cause of radiation.

It might be very briefly stated that when any form becomes radioactive, certain conditions have been fulfilled and certain results brought about, which conditions and results might be summed up as follows:

The radioactive form is one which has run through its appointed cycles, through its wheel of life, great or small, which has been turned with adequate frequency, so that the volatile life-essence is ready to escape from that form and merge itself in the greater form of which the lesser is but a part. It must be remembered in this connection that radiation occurs when the etheric or true form becomes responsive to certain types of force. Radiation, as it is occultly understood, does not concern itself with the escape from the physical or dense form, but with that period in the life of any living entity (atomic, human or divine) wherein the etheric or pranic body is in such a state that it can no longer limit or confine the indwelling life.

Radiation comes about when the internal, self-sufficient life of any atom is offset by a stronger urge, or pull, emanating from the enveloping greater existence of whose body it may form a part. This is nevertheless only true when it is caused by the *pull upon the essential life by the essential life of the greater form;* it is not due to the attractive power of the form aspect of that greater life. A very definite distinction must here be made. It is the failure to recognise this that has led so many alchemical students and scientific investigators to lose their way, and thus negate the conclusions of years of study. They confuse the impulse of the atom to respond to the vibratory magnetic pull of the more powerful and comprehensive form with the true esoteric attraction which alone produces "occult radiation,"—that of the central essential life of the form in which the element under consideration may have place. It is very necessary to make this clear from the start. Perhaps the whole subject may be clearer if we consider it in the following way.

The atom in a form revolves upon its own axis, follows its own revolution, and lives its own internal life. This concerns its primary awareness. As time progresses it becomes magnetically aware of the attractive nature of that which envelops it on all sides, and becomes conscious of the form which surrounds it. This is its secondary awareness but it still concerns what we might, for lack of a better term, call matter. The atom, therefore, has an interplay with other atoms.

Later, the atom in a form becomes aware that it not only revolves upon its axis, but that it also follows an orbit around a greater centre of force within a greater form. This is *tertiary* awareness, and is caused by the magnetic pull of the greater centre being felt, thus causing an urge within the atom which impels it to move within certain specific cycles. This awareness, esoterically understood, concerns itself with substance or with the true form within the objective form.

Finally, the attractive pull of the greater centre becomes so powerful that the positive life within the atom (whatever type of atom it may be and in whatever kingdom) feels the force of the central energy which holds it, along with other atoms, coherently fulfilling their function. This energy penetrates through the ring-pass-not, evokes no response from what might be called the electronic or negative lives within the atomic periphery, but does evoke a response from the essential, positive nucleus of the atom. This is due to the fact that the essential life of any atom, its highest positive aspect, is ever of the same nature as that of the greater life which is drawing it to itself. When this is felt sufficiently strongly, the atomic cycle is completed, the dense form is dispelled, the true form is dissipated, and the central life escapes to find its greater magnetic focal point.

Through this process (which is found throughout the solar system in all its departments) every atom in turn becomes an electron. The positive life of any atom in due course of evolution becomes negative to a greater life toward which it is impelled or drawn, and thus the process of evolution carries every life invariably through the four stages enumerated above. Within the three lower kingdoms of nature, the process is undergone unconsciously, according to the human connotation of that term; it is consciously passed through in the human kingdom, and in the higher spheres of existence, with an enveloping consciousness which can only be hinted at in the ambiguous term "self-conscious group realisation."

A Treatise on Cosmic Fire, pp. 1063/5

• In studying the subject of radiatory activity, we are dealing with the effect produced by the inner essence as it makes its presence felt through the form, when the form has been brought to a stage of such refinement that it becomes possible.

When this realisation is applied to all the forms in all the kingdoms, it will be found possible to bridge the gaps existing between the different forms of life, and the "elements" in every kingdom, and those unifying radiating centres will be found. The word "element" is yet confined to the basic substances in what is called essential matter, and the chemist and physicist are busy with such lives; but their correspondence (in the occult sense of the term) is to be found in every kingdom in nature, and there are forms of life in the vegetable kingdom which are occultly regarded as "radioactive," the eucalyptus tree being one such form. There are forms of animal life equally at an analogous stage and the human unit (as it approaches "liberation") demonstrates a similar phenomenon.

Again, as a planetary scheme nears its consummation, it becomes "radioactive," and through radiation transfers its essence to another "absorbent planet," or planets, as is the case with a solar system also. Its essence, or true Life, is absorbed by a receiving constellation, and the outer "case" returns to its original unorganised condition.

A Treatise on Cosmic Fire, p. 1062

• (Published: 1925) In the tremendous event which is impending, in the great revelation which is near at hand, the Hierarchy will again take advantage of the time and the energy to bring about certain events which will work out primarily in the human kingdom but which will also be seen as force regeneration in *the mineral kingdom.* The energy, when first felt in the human kingdom, brought about the conditions which caused the tremendous activity which resulted in war, and which is causing the present world stress; in the mineral kingdom it affected certain of the minerals and elements, and the radioactive substances made their appearance. This characteristic (or radioactivity) of pitchblende and the other involved units is comparatively a new development under the evolutionary law, and one which, though latent, only needed the drawing forth of the type of energy now beginning to pour in on the earth. This force began to flow in at the end of the eighteenth century, and its full effect is by no means yet felt, for it will be several hundred years before it passes away. By means of it, certain discoveries are possible, and the new order comes in upon it. The Great Ones, Who know the time and the hour, will bring about, in our rootrace, that which corresponds to the occurrences in the earlier third and fourth races. *A Treatise on Cosmic Fire*, pp. 716/7

• All atoms become radioactive as the result of a response to a stronger magnetic centre which response is brought about through the gradual evolutionary development of consciousness of some kind or another. This is known to be true in a small degree in connection with the mineral kingdom though scientists have not yet admitted that radiation is thus caused. Later they will, but only when this general theory which is here laid down in connection with all atoms is admitted by them to be a plausible hypothesis. Then the goal of their endeavour will be somewhat changed; they will seek to ascertain through clear thinking and a study of the involved analogy what focal points of magnetic energy may be regarded as existing, and how they affect the atoms in their environment. One hint only can here be given. Light upon these dark problems will come along two lines.

First, it will come through the study of the place of the solar system in the universal whole, and the effect that certain constellations have upon it; secondly, it will come through a close study of the effect of one planetary scheme upon another, and the place of the moon in our own planetary life. This will lead to a close investigation of polar conditions in the earth, of the planetary magnetic currents, and of the electrical intercourse between our earth, and the Venusian and Martian planetary schemes. When this has been accomplished, astronomy and esoteric astrology will be revolutionised, and the nature of solar energy as an expression of an Entity of the fourth rank will be appreciated. This will come at the close of this century after a scientific discovery of even greater importance to the scientific world than that as to the nature of the atom. Until that time it will be as difficult to express the hylozoistic conception in terms of exact science as it would be for the sixteenth century ancestor of present humanity to conceive of the atom as being simply an aspect of force, and not objective and tangible. Hence further elucidation will but serve to confuse.

A Treatise on Cosmic Fire, pp. 1069/70

• We have seen how in this question of the transference of the life from form to form, the work proceeds under rule and order, and is effected through the co-operation of the devas in the first instance, and the application of external agents to the atom or form involved, and in the second place (involving the most important and lengthy stage of the procedure) through the subsequent reaction within the atom itself, which produces an intensification of the positive burning centre, and the consequent escape (through radioactivity) of the volatile essence.

At all the different stages, the fire elementals perform their part, aided by the fire devas who are the controlling agents. This is so on all the planes which primarily concern us in the three worlds—different groups of devas coming into action according to the nature of the form concerned, and the plane on which the transmutation is to take place. Electric fire passes from atom to atom according to law, and "fire by friction" responds, being the latent fire of the atom, or its negative aspect; the process is carried on through the medium of solar fire, and herein lies the secret of transmutation and its most mysterious angle. Fire by friction, the negative electricity of substance, has been for some time the subject of the attention of exoteric science, and investigation of the nature of positive electricity has become possible through the discovery of radium.

Keely, as H. P. B. hinted, had gone far along this path, and knew even more than he gave out, and others have approached, or are approaching, the same objective. The next step ahead for science lies in this direction, and should concern the potential force of the atom itself, and its harnessing for the use of man. This will let loose upon earth a stupendous amount of energy. Nevertheless, it is only when the third factor is comprehended, and science admits the agency of mental fire as embodied in certain groups of devas, that the force of energy that is triple, and yet one in the three worlds, will become available for the helping of man. This lies as yet far ahead, and will only become possible towards the end of this round; and these potent forces will not be fully utilised, nor fully known till the middle of the next round. At that time, much energy will become available through the removal of all that obstructs. This is effected, in relation to man, at the Judgment separation, but it will produce results in the other kingdoms of nature also. A portion of the animal kingdom will enter into a temporary obscuration, thus releasing energy for the use of the remaining percentage, and producing results such as are hinted at by the prophet of Israel when he speaks of "the wolf lying down with the lamb"; his comment "a little child shall lead them" is largely the esoteric enunciation of the fact that three fifths of the human family will stand upon the Path, 'a little child' being the name applied to probationers and disciples. In the vegetable and mineral kingdoms a corresponding demonstration will ensue, but of such a nature as to be too obscure for our comprehension.

The central factor of solar fire in the work of transmutation will come to be understood through the study of the fire devas and elementals, who are fire, and who are, in themselves (essentially and through active magnetic radiation), the external heat or vibration which produces:

    The force which plays upon the spheroidal wall of the atom.
    The response within the atom which produces radiation or the escape of volatile essence.

Speaking cosmically, and regarding the solar system as itself a cosmic atom, we would consider that:

    The abstractions or entities who indwell the form are "electric fire."
    The material substance which is enclosed within the ring-pass-not viewing it as a homogeneous whole, is "fire by friction."
    The fire devas from the cosmic mental plane (of whom Agni and Indra are the embodiers along with one whose name is not to

be given) are the external agencies who carry on cosmic transmutation.

This triple statement can be applied to a scheme, a chain, or a globe also, remembering ever that in connection with man the fire which is his third aspect emanates from the systemic mental.

A Treatise on Cosmic Fire, pp. 491/4

• Only when the fires of matter are blazing brightly, and become radiatory, does it become possible for the fire of mind to show forth, even though ever inherently present. Only when these two fires of matter and of mind have reached a stage of energetic heat and light, can the electric fire of Spirit show forth in its glory. Only again when these three are unitedly burning does the fire of matter die down for lack of that which it may consume, and only when that occurs is it possible for the fires of mind (on mental levels) to burn up that which it has hitherto animated. When this is accomplished, the fire of pure Spirit (increased and intensified by the gaseous essence of the fire of matter, or "fire by friction," and coloured, and rendered radiatory by the fire of mind) blazes forth in perfected glory, so that naught is seen save one vibrant flame. This idea can be extended away from Man to a Heavenly Man, and again to the Logos in His cosmic relationship.

A Treatise on Cosmic Fire, p. 731

1. TRANSMUTATION IN THE HUMAN KINGDOM

• The causal body is then (expressed in terms of fire) a blazing centre of heat, radiating to its group warmth and vitality. Within the periphery of the egoic wheel can be seen the nine spokes rotating with intense rapidity and—after the third Initiation—becoming fourth dimensional, or the wheels "turn upon" themselves. In the midst forming a certain geometrical triangle (differing according to the ray of the Monad) can be seen three points of fire, or the permanent atoms and the mental unit, in all their glory; at the centre can be seen a central blaze of glory growing in intensity as the three inner petals respond to the stimulation. *When the fire of matter, of "fire by friction," becomes sufficiently intense; when the fire of mind of solar fire (which vitalises the nine petals) becomes equally fierce, and when the electric spark at the innermost centre blazes out and can be seen, the entire causal body becomes radioactive.* Then the fires of substance (the vitality of the permanent atoms) escape from the atomic spheres, and add their quota to the great sphere in which they are contained; the fire of mind blends with its emanating source, and the central life escapes. This is the great liberation. The man, in terms of human endeavour, has achieved his goal. He has passed through the three Halls and in each has transferred that which he gained therein to the content of his consciousness; he has in ordered sequence developed and opened the petals of the lotus—first opening the lower three, which involves a process covering a vast period of time. Then the second series of petals are opened, during a period of time covering his participation intelligently in world affairs until he enters the spiritual kingdom at the first Initiation; and a final and briefer period wherein the three higher or inner ring of petals are developed and opened.

A Treatise on Cosmic Fire, pp. 542/3

2. TRANSMUTATION IN THE MINERAL KINGDOM

• The mineral kingdom will find its highest manifestation in matter of the fourth ether, and this transmutation is already taking place, for all the radioactive substances now being discovered are literally becoming matter of the fourth ether. The mineral kingdom is *relatively* nearing its possible manvantaric perfection, and by the time the seventh round is reached all mineral lives (not forms) will have been transferred to another planet. This will not be so with the other three kingdoms. *A Treatise on Cosmic Fire,* pp. 935/6

• . . . the note of the mineral kingdom is the basic note of substance itself, and it is largely the sounding of the note combinations, based on this key, which produces the great world cataclysms, wrought through volcanic action. Every volcano is sounding forth this note, and, for those who can see, the sound and colour (occultly understood) of a volcano are a truly marvellous thing. Every gradation of that note is to be found in the mineral kingdom which is itself divided into three main kingdoms:

a. The baser metals, such as lead and iron, with all allied minerals.
b. The standard metals, such as gold and silver, which play such a vital part in the life of the race, and are the mineral manifestation of the second aspect.
c. The crystals and precious stones, the first aspect as it works out in the mineral kingdom—the consummation of the work of the mineral devas, and the product of their untiring efforts.

When scientists fully appreciate what it is which causes the difference between the sapphire and the ruby, they will have found out what constitutes one of the stages of the transmutative process, and this they cannot do until the fourth ether is controlled, and its secret discovered. As time progresses, the transmutation, for instance, of coal into diamonds, of lead into silver, or of certain metals into gold, will hold no appeal for man, for it will be recognised that the outcome of such action would cause deterioration of the standard, and result in poverty instead of the acquirement of riches; man will eventually come to the realisation that in atomic energy, harnessed to his need, or in the inducing of increased radioactivity, lies for him the path to prosperity and riches. He will, therefore concentrate his attention on this higher form of life transference and

a. Through knowledge of the devas,
b. Through external pressure and vibration,
c. Through internal stimulation,
d. Through colour applied in stimulation and vitalisation,
e. Through mantric sounds

he will find the secret of atomic energy, latent in the mineral kingdom, and will bend that inconceivable power and force to the solution of the problems of existence. Only when atomic energy is better understood and the nature of the fourth ether somewhat comprehended, shall we see that control of the air which lies inevitably ahead.

A Treatise on Cosmic Fire, pp. 495/6

• Again speaking symbolically . . . the mineral kingdom marks the point of unique condensation. This is produced under the action of fire and by the pressure of the "divine idea". Esoterically speaking, we have, in the mineral world, the divine Plan hidden in the geometry of a crystal, and God's radiant beauty stored in the colour of a precious stone. In miniature and at the lowest point of manifestation, we find the divine concepts working out. The goal of the universal concept is seen when the jewel rays forth its beauty, and when radium sends forth its rays, both destructive and constructive. If you could really understand the history of a crystal, you would enter into the glory of God. If you could enter into the attractive and the repulsive consciousness of a piece of iron or lead, you would see revealed the complete story of evolution. If you could study the hidden processes which go on under the influence of fire, you would enter into the secret of initiation. When the day comes when the history of the mineral kingdom can be grasped by the illumined seer, he will then see the long road that the diamond has travelled, and—by analogy—the long road that all sons of God traverse, governed by the same laws and unfolding the same consciousness. *Esoteric Psychology* I, pp. 226/7

3. ALCHEMY

• It was in connection with this transmutative process that the alchemists of old occupied themselves, but seldom did they reach the stage wherein it was possible for them to concern themselves with the response of the two types of positive energy to each other, and with the consequent escape of a lesser positive force to its greater attractive centre. When they did (with a few exceptions) they were brought up against a dead wall, for though they had succeeded in locating the radiating principle in substance, or in the true form, and had managed to pierce through (or to negate) both the dense physical body and the etheric form, yet they had no perception of the nature of the central force which was drawing the life they were concerned with out of its apparently legitimate sphere into a new realm of activity. Some few did possess this knowledge but (realising the danger of their conclusions) refused to put in writing the result of their investigations.

If students will study the laws of transmutation, as already apprehended, and above all, as incorporated in the writings of Hermes Trismegistus, bearing this in mind, some interesting results might be brought about. Let them remember that that which "seeks liberty" is the central electric spark; that this liberty is achieved first of all through the results brought about by the activity of the "frictional fire" which speeds up its internal vibration; then by the work upon the atom, or the substance of solar fire, which causes:

a. Orbital progression,
b. Stimulative vibration,
c. Awakened internal response,

until finally electric fire is contacted. This is true of all atoms:
a. The atom of substance,
b. The atom of a form whatsoever it be,
c. The atom of a kingdom in nature,
d. The atom of a planet,
e. The atom of a solar system.

In every case the three fires or types of energy play their part; in every case the four stages are passed through; in every case transmutation, transference, or radiation takes place, and the result of the escape of the central positive energy is achieved, and its absorption into a greater form, to be held in place for a specific cycle by the stronger energy.

This process of rendering radioactive all the elements has, as we have seen, occupied students down the ages. The alchemists of the middle ages beginning with the simpler elements and starting with the mineral kingdom sought to find out the secret of the liberating process, to know the method of release, and to understand the laws of transmutation. They did not succeed in the majority of cases because, having located the essence, they had no idea how to deal with it when released, nor (as we have seen) had they any conception as to the magnetic force which was drawing the released essence to itself.

RADIATION AND TRANSMUTATION

To comprehend the law and therefore to be able to work perfectly with it, the experimenting student must have the ability to release the essence from its form. He must know the formulae and words which will direct it to that particular focal point in the mineral kingdom which stands in the same correspondential relation to the mineral monad as the Ego on its own plane stands to the man who casts off his physical and true forms through death. This involves knowledge only committed to the pledged disciple; if chance students stumble upon the law, and theoretically know the process, they would do well to proceed no further until they have learned how to protect themselves from the interplay of forces. As we well know, the workers with radium, and those who experiment in the world's laboratories, suffer frequently from loss of limb or life; this is due to their ignorance of the forces they are dealing with. The liberated essences become conductors of the greater force which is their magnetic centre, because they are responsive to it, and it is this force which produces the distressing conditions sometimes present in connection with radioactive substances. Every radioactive atom becomes, through this conductive faculty, a releasing agent; and they consequently cause what we call burns. These burns are the result of the process of releasing the essential life of the atom of physical substance being dealt with.

A Treatise on Cosmic Fire, pp. 1065/8

• We have dealt in broad and general manner with this question of electricity and have seen that fire essence or substance is resolved through internal activity and external heat in such a manner that the electric fire at the centre of the atom is liberated and seeks a new form. This is the aim of the transmutative process and the fact that hitherto alchemists working in the mineral kingdom have failed to achieve their objective has been due to three things:

First. Inability to contact the central electric spark. This is due to ignorance of certain of the laws of electricity, and above all, ignorance of the set formula which covers the range of the electrical influence of that spark.

Second. Inability to create the necessary channel or "path" along which the escaping life may travel into its new form. Many have succeeded in breaking the form so that the life has escaped but they have not known how to harness or guide it and all their labour has consequently been lost.

Third. Inability to control the fire elementals who are the external fire which affects that central spark through the medium of its environment. This inability is especially distinctive of the alchemists of the fifth root race who have been practically incapable of this control, having lost the Words, the formulas, and the sounds. This is the consequence of undue success in Atlantean days, when the alchemists of the time, through colour and sound so entirely controlled the elementals that they utilised them for their own selfish ends and along lines of endeavours outside their legitimate province. This knowledge of formulas and sounds can be comparatively easily acquired when man has developed the inner spiritual ear. When this is the case, the transmutative processes of the grosser kind (such as are involved in the manufacture of pure gold) will interest him not at all and only those subtler forms of activity which are connected with the transference of life from graded form to form will occupy his attention. *A Treatise on Cosmic Fire,* pp. 494/5

- As time proceeds, man will gradually do four things:

1. Recover past knowledge and powers developed in Atlantean days.
2. Produce bodies resistant to the fire elementals of the lower kind which work in the mineral kingdom.
3. Comprehend the inner meaning of radioactivity, or the setting loose of the power inherent in all elements and all atoms of chemistry, and in all true minerals.
4. Reduce the formulas of the coming chemists and scientists to SOUND, and not simply formulate through experiment on

paper. In this last statement lies (for those who can perceive) the most illuminating hint that it has been possible as yet to impart on this matter.

A Treatise on Cosmic Fire, p. 486

• Indication only is possible; it is not permissible here to give out the transmutative formulas, or the mantrams that manipulate the matter of space. Only the way can be pointed to those who are ready, or who are recovering old knowledge (gained through approach to the Path, or latent through experience undergone in Atlantean days) and the landmarks indicated hold sufficient guidance to enable them to penetrate deeper into the arcana of knowledge. The danger consists in the very fact that the whole matter of transmutation concerns the material form, and deva substance. Man, being not yet master even of the substance of his own sheaths, nor in vibratory control of his third aspect, incurs risk when he concentrates his attention on the Not-Self. It can only be safely done when the magician knows five things:

1. The nature of the atom.
2. The keynote of the planes.
3. The method of working from the egoic level through conscious control, knowledge of the protective sounds and formulas, and pure altruistic endeavour.
4. The interaction of the three fires, the lunar words, the solar words, and later a cosmic word.
5. The secret of electrical vibration, which is only realised in an elementary way when a man knows the keynote of his own planetary Logos.

All this knowledge as it concerns the three worlds is in the hands of the Masters of the Wisdom, and enables Them to work along the lines of energy or force, and not with what is usually understood when the word 'substance' is used. They work with electrical energy, concerning Themselves with positive electricity, or with the energy of the positive nucleus of force within the atom, whether it is the atom of chemistry, for instance, or the human atom. They *deal with the soul of things*. The black magician works with the negative aspect, with the electrons, if I might so term it, with the sheath, and not with the soul. This distinction must be clearly borne in mind. It holds the clue to the non-interference of the whole Brotherhood in material matters and affairs, and Their concentration upon the force aspect, upon the centres of energy. *They reach the whole through the agency of the few centres in a form.* *A Treatise on Cosmic Fire*, pp. 481/3

XV. IMMORTALITY

• *Letter to a Scientist* - February, 1944

My brother:

. . . The discoveries of science are as yet inadequate for the fulfillment of the prophecies I made in *A Treatise on the Seven Rays.* Towards the close of this century and when the world situation has clarified and the period of reconstruction is drawing to a close, discoveries will be made which will reveal some hitherto unrealised electrical potencies. I know not what other word to use for these electrical rays which will make their presence felt and lead to possibilities beyond the dreams of investigators today. The coming science of electricity will be as different next century as the modern usages of electricity differ from the understanding of the Victorian scientist.

In connection with your query anent the photography which concerns itself with departed souls, I would advise you that understanding of process will come from a study of the—photographing of thoughtforms. A beginning was made in this connection by the great French scientist, d'Arsonval, of Paris. A.A.B. can tell you something of this if you do not already know. Light on the subject will come through this, through the perfecting of the plates of reception and their greatly increased sensitivity, and through the relating of electricity to photography. You may deem it well-nigh impossible to make plates of much greater sensitivity than those in use in the best equipped laboratories. But this is not so. Along this line of thought-photography and electrical equipment, will come the solution. It is the thought of those on the other side, and their ability to project thoughtforms of themselves, plus the providing of adequately sensitive plates or their equivalent, which will mark a new era in so-called "spirit photography". People frequently are so preoccupied with the tangible instrument on this side of the veil that they neglect the factor of what must be contributed from the other side by those who have passed over.

The work will be done from there, with the material aid which as yet has not been provided in the outer scientific field.

To bring this about, collaboration of a conscious medium (not a trance medium, but someone who is consciously clairvoyant and clairaudient) will be required. There are many such growing up among the children of today, and the next generation after them will provide still more. The separating veil will disappear through the testimony of the thousands of those who can see phenomena and hear sounds which lie outside the range of the tangible.

You say that the spirits state that they cannot stand electricity. What is meant is that they cannot stand electricity as it is at present applied. This is an instance of the inaccurate statements passed on by ignorant mediums or by those who on the other side have no more understanding of the laws of electricity than they probably had in the physical body. There is nothing but electricity in manifestation, the "mystery of electricity" to which H.P.B. referred in to in *The Secret Doctrine*. Everything in Nature is electrical in nature; life itself is electricity, but all that we have contacted and used today is that which is only physical and related to and inherent in the physical and etheric matter of all forms.

It must be remembered that the so-called "spirits" are functioning in the illusory astral body, while advanced "spirits" are only functioning as minds, and can therefore be reached solely by minds and in no other way. It will never be possible to photograph the mental vehicle; only the astral body will be susceptible of photographing. The grosser the person in the body, desire and appetite, the more easily will he be photographed after passing over (if anyone wants to photograph him!), and the more advanced the person, the more difficult it will be to get a photograph.

As regards the use of radio as a means of communication with the "spirit world," the present electrical instruments are too slow in vibratory activity (if I may use such an unscientific term) to do the work; if astrally clothed "spirits" approach them they are apt to have a shattering effect. Yet the first demonstration of existence after death, in such a way that it can be registered upon the physical plane, will come via the radio, because sound always precedes vision. Think on this.

However, no radio now exists which is sufficiently sensitive to carry sound waves from the astral plane.

Future scientific discoveries, therefore, hold the secret. This is no evasion on my part, but a simple statement of fact. Electrical discovery is only in the initial stage and all that we have is simply a prelude to the real discovery. The magic of the radio would be completely unbelievable to the man of the eighteenth century. The discoveries and developments lying ahead in the twenty-first century will be equally unbelievable to the man of this century. A great discovery in relation to the use of light by the power and the directive agency of thought will come at the end of this century or the beginning of the next. Two small children—one living in this country (U.S.A.) and one in India—will work out a formula along scientific lines which will fill in some of the existing gaps in the scale of light vibration, carrying on from the high frequency rays and waves as you now have them. This will necessitate instruments hitherto undreamt of but really quite possible. They will be so sensitive that they will be set in motion by the power of the human eye under the focussed direction of thought. From then on tangible rapport with the spirit world will be possible. I cannot do more than give you the clue.

. . . An ordinary treatise on electricity such as is studied by electrical engineers would have been completely incomprehensible to even the most highly educated man two hundred years ago, or even one hundred, and so it is now. In the meantime, work with thought photography as a prelude to the coming science, for out of that and the gradual development of more sensitive modes of registering and recording subtle phenomena will come the new idea and possibilities. Does it mean anything to you when I say that electricity and photography are closely related because the human being is electrical in origin and nature? This must be demonstrated on the physical plane by the aid of the needed sensitive apparatus.

Esoteric Healing, pp. 376/9

• THE PROBLEM OF IMMORTALITY

The third area of doubt,—doubt as to the fact of immortality—

will be solved before long in the realm of science, as the result of scientific investigation. Certain scientists will accept the hypothesis of immortality as a working basis upon which to base their search, and they will enter upon that search with a willingness to learn, a readiness to accept and a desire to formulate conclusions based upon reiterated evidence. These conclusions will, in their turn, form the basis for another hypothesis. Within the next few years the fact of persistence and of the eternity of existence will have advanced out of the realm of questioning into the realm of certainty. The problem will have shifted further back. There will be no question in anyone's mind that the discarding of the physical body will leave a man still a conscious living entity. He will be known to be perpetuating his existence in a realm lying behind the physical. He will be known to be still alive, awake and aware. This fact will be demonstrated in several ways. The development of a power within the physical eye of a human being (a power which has always been there, but which has been very little used) will reveal the etheric body, the "double," as it is sometimes called; and men will be seen occupying that body in some definite spatial area whilst their dead or disintegrating physical body has been left behind. Then again, the growth in the number of those people who have the power to use the "single eye," sometimes called the "reawakened third eye," will also add to the demonstration of the truth of immortality, for they will with facility see the man who has discarded his etheric body as well as his physical body. By the very weight of their numbers, and by the reputability of their position, they will carry their point. Through a discovery also in the field of photography, now being investigated, will the fact of survival be proven. Through the use of the radio by those who have passed over will communication be eventually set up, and reduced to a true science.

Nevertheless, certain imminent happenings will do more to annihilate the veil between the seen and the unseen than any other line of activity hitherto initiated. Of this I may not speak beyond telling you that an illumination will be set up and a radiance revealed which will result in a tremendous stimulation of mankind and bring about an awakening of a new order. Man will be keyed up to a perception and

IMMORTALITY 289

to a contact which will enable him to *see through,* which will reveal the nature of the fourth dimension, and will blend the subjective and the objective together into a new world. Death will lose its terrors, and that particular fear will come to an end.

Men are so occupied with their demand for light, so earnest in their cry for release from the present blindness, and so anxious for relief from the surrounding chaos, that they are apt to forget that from the inner side there is also a great effort and "push" to help, on the part of the Custodians of the Plan and Their assistants. This urge on Their part to help is more active than ever before, as human beings demand more potently the privilege of light. A demand from the race, plus a response from the waiting Hierarchy, must inevitably produce potent results. The urge to know and the urge to teach are assuredly related and a part of the natural process of conscious development. The next few decades will mark a happening of such profound and widespread consequences that the present era in which we live will come to be looked upon as the dark ages. Science will penetrate deeper into the realm of the intangible, and work in mediums and with apparatus hitherto unknown. The release of the potencies in an atom will mark a revolutionary era, and science will have much to discard and much to give as it works with energies and forms of life hitherto unrecognised. The spiritualists will make a discovery whereby the means of contact with those who function out of the physical body will be greatly facilitated, and a group of mediums will begin to act as intermediaries for a number of scientists on the inner side of life and those who are still in physical bodies. Through the activity of the real esoteric schools, a technique of training will be instituted which will develop the new powers that will substantiate the old truth and turn men's beliefs into certainties. Through the stimulating and occultly scientific work of the department of religions, men will come to new knowledge and awareness, and will arrive at an uplift that will bring mankind to the Mount of Transfiguration. Through the work of the department of government, men will come to an understanding of those ideas which are needed to carry the nations the next step forward to mutual help.

I shall try to express the deepest objective of the Brotherhood, so that you can understand and cooperate. Humanity is intended to act as a power house through which certain types of divine energy can flow to the various forms of life found in the subhuman kingdoms. This flow of energy must be intelligently apprehended and intelligently directed, and thus will be brought to an end conditions of decay and of death now prevalent everywhere. Thus mankind can link the higher and the lower manifestations of Life, but this will be possible only when men themselves have (within themselves) linked their higher and their lower aspects. This is, and should be, one of the objectives of all esoteric training. Men are intended to acquire the facility to function freely in either direction, and so with ease contact the life of God as it flows through those forms we call superhuman, and those which are subhuman. Such is the emerging goal.

Esoteric Psychology I, pp. 183/6

ARCANE SCHOOL TRAINING

Training for new age discipleship is provided by the *Arcane School.* The principles of the Ageless Wisdom are presented through esoteric meditation, study and service as a *way of life.*

****

www.lucistrust.org/arcaneschool